mills
grammar
school
framlingham

Harvey Form Prize,

Heulwen Jones 1979

INSECTS
AND OTHER INVERTEBRATES
IN COLOUR

Åke Sandhall

INSECTS
AND OTHER INVERTEBRATES
IN COLOUR

*Adapted and revised from the
Swedish edition by Ronald M. Dobson*

Lutterworth Press · Guildford and London

Colour Photographs:
Anders Ardö nos. 158, 221.
Ingmar Holmåsen nos 81, 127, 152,
171, 214, 215, 216, 219,
230, 235, 240, 273, 274, 419.
Åke Sandhall all others.
Drawings:
Mats Lind, Ronald M. Dobson.
Text:
Åke Sandhall, Birgit Sandhall, Ronald M. Dobson.
The text has been checked by 22 specialists
in different groups.
Species determinations:
Hugo Andersson, Christine Dahl,
Per Douwes, Carl H. Lindroth,
Ulf Norling, Arne Samuelsson and others.
Photographic equipment for most colour plates:
Cameras: Canon FX, Canon FTb.
Lenses: Canon Macro FL 50 mm F3.5,
Canon FLM 100 mm F4, Canon FD 300 mm F5.6.
Canon bellows FL, Canon extension rings FL.
Electronic flash: Braun F 655, Minicam ringflash,
specially made telescopic extension arm.
Film: Kodachrome II.

First published in Great Britain 1975

ISBN 0 7188 2214 5

Contents

Preface

The aim of this book is to introduce the reader to the world of the insects and other beautiful and fascinating small invertebrate animals which are of vital importance to man and his environment today. The main part of the work consists of 466 colour photographs of selected species taken in their natural settings and each of these is accompanied by a brief commentary. I hope that the basic information thus provided will stimulate interest in, and encourage the more thorough study of, these small animals. A brief illustrated review of the animal kingdom, an annotated bibliography and a glossary of technical terms have, therefore, been provided to assist with such extension studies.

The selection of species included in an introductory work such as this will always be a matter for controversy and it is not easy to choose a mere 400 or so out of an available total of some 30,000. In making my choice I have attempted to obtain as wide a coverage of different animal groups as possible but at the same time have given preference to common and easily observed species, especially those with distinctive colouration or with interesting life-histories or habits. Animals of economic significance such as agricultural and household pests are also well represented. My final choice was influenced by the short time available, a bare three summers.

I am deeply grateful to all who helped me with this work, in particular to my family, to Prof. Carl H. Lindroth and to my friends in the Entomological Society of Lund.

Åke Sandhall

How to Use the Book

The colour plates and commentaries are grouped into 11 sections each of which deals with a selection of animals usually associated with a particular type of environmental setting. If an unfamiliar species is found in any one of these settings it should be possible to identify it *at least partially* by reference to the colour photographs of the appropriate section or, if necessary, to those of other sections dealing with rather similar environments e.g. a beetle found on the surface of a pond would first be sought in section 8, 'On and in Freshwater', but it could well appear in section 7, 'By lake and stream'. While in general these environmentally arranged sections form a useful guide to distribution it should be noted that some species with considerable powers of dispersal often appear far from their 'usual' settings. The Large white butterfly, *Pieris brassicae* (Nos. 52, 53), for example, while mainly associated with cultivated plants is a migratory species and may be found almost anywhere.

The book is not, of course, primarily intended as an identification manual; only about 1% of the British invertebrate fauna is portrayed and in any case animals such as these cannot usually be identified by photographs alone. However the selection of species has been such that many of those included are of distinctive appearance and can be named from the pictures. Moreover as most of them are common and familiar forms (many have English common names) the chances that any particular specimen can be named are relatively high.

If a specimen cannot be satisfactorily identified, the reader should not be discouraged because by careful use of the pictures and the line illustrations in **the systematic survey** he will usually be able to assign it to its correct zoological group. Thereafter the identification can be completed by referring to one of the advanced texts detailed in **the bibliography.** With practice familiarity with the various groups will soon be gained. **The glossary** explains most of the technical terms used in this book and likely to be encountered in the more advanced texts and **the index** lists the common and scientific names of both species and higher zoological groups.

The following information accompanies each plate:
1. **The English common name** of the animal or of its group. Its stage if immature (larva, pupa etc.) and sometimes the sex (δ = male; $\female$ = female) are indicated.
2. **The Latin name** – italicised in accordance with convention.
3. **Its size** in metric units. L = body length of adult or full-grown immature (projecting parts such as antennae and cerci excluded), Ws = span of the fully opened wings (moths etc), H = height and D = diameter of tall and spherical snail shells respectively.
4. **The time of year** when the animal normally occurs in its natural setting. This can only be approximate because variations occur with latitude. In the North the period of summer activity is usually shorter than in the South.
5. **Abundance.** 'Common' species occur

in suitable environments almost throughout the country whereas 'local' species occupy only a proportion of available sites. 'Rare' species are either extremely localised or occur only sporadically.

6. Additional notes on **environmental associations.**
7. **Selected information** on life-history, feeding habits, distribution, economic significance and folklore.

The Pronunciation of Latin Names

In Britain, the pronunciation of Latin names closely resembles that of English words of similar construction. Both short vowels (e.g. ă) and long ones (e.g. ā) occur and are pronounced as in the following English words hăt, māte; nĕt, ēqual; tĭn, īron; hŏt, rōse; hŭt, ūse. Examples, with the syllables separated are
Cī'mĕx, Nā'bĭs, Crămb'ŭs, Pĭss'ō'dēs and *Ū'rŏc'ĕr'ŭs.*
The letter Y can be pronounced ĭ as in *Ĕr'ўnn'ĭs* or ī as in *Cȳ'clōps.*

Paired vowels may be sounded separately as in *Ăg'rĭ'ōt'ēs;* one of the pair only may be heard, e.g. ae & oe are usually pronounced ē, eu and ue are usually ū; or the pair may have a distinct sound of its own, e.g. au = aw as in raw, oi = oy as in toy. Very occasionally a printed diaresis (··) is used over the second of a pair of vowels; this indicates that the two vowels are pronounced separately, e.g. *Aëdes, Meloë.* C is pronounced s before a soft vowel or soft vowel sound (i, y, ae, oe and sometimes e and eu) and as k before a hard vowel (a, o and sometimes e) or before a consonant. G is usually hard as in 'good', but common usage often softens it to j e.g. in Geometridae it is pronounced as in 'geometry'. When g immediately precedes n in the same syllable it is silent as in 'gnash'. Ch is always k as in choir; sch should be sk but often becomes sh and ph is always f.

Two syllable words have a stress on the first syllable. In longer words the length of the penultimate syllable determines the position of the stress; if this is long it is stressed, if short, the third syllable from the end bears the stress. A syllable is long when it contains a long vowel or a diphthong and when its vowel is followed by at least two consonants. In this connection ch, ph, rh and th count as single consonants.

1 Swallow-tail
Papilio machaon
Ws 58–88 mm April–Aug Flowery meadows, woods and near water

This large, graceful, swift-flying butterfly occurs throughout continental Europe. In Britain, however, breeding populations are found in only a few parts of the Norfolk fens where the larvae of a distinct sub-species feed chiefly on milk parsley. The female is generally larger than the male and lays greenish, spherical eggs on the lower surface of leaves of umbelliferous plants. There is one generation per year in N. Europe but in the South 2–3 may occur. Winter is spent as a pupa. Caterpillar No 100.

2 Apollo
Parnassius apollo
Ws 62–92 mm July–Aug Mainly on mountains

This butterfly does not occur in Britain but inhabits most mountainous districts of Europe at altitudes of 2,500–6,000 ft. In the North, however, it descends lower and may be found near coastal cliffs. The caterpillar is black with grey-blue tubercles on its back and reddish-yellow spots on its sides. It feeds on plants of the stone-crop family. The pupa lies on the ground in a loosely woven cocoon. It is dark and is covered with bluish-grey wax. Winter is usually spent as an egg.

3 Brimstone ♂
Gonepteryx rhamni
Ws 55–60 mm March–Sept Common; on flowers, mainly in wooded areas

The photograph shows the male; the female is pale green. These butterflies emerge in August and, as winter approaches, seek shelter for hibernation. Activity is resumed in the first sunny days of spring, the males appearing before the females. Bottle-shaped, longitudinally-ribbed, upright, greenish eggs are laid individually on the under-sides of leaves of buckthorn, often along the midrib. The caterpillar, when mature in July, attaches itself to a leaf by means of a silken thread and pupates. The pupa resembles a curled leaf.

4 Orange Tip ♂
Anthocaris cardamines
Ws 31–50 mm　May–June
Common; meadows and glades

This species hibernates as a pupa
attached to the stalk of a host plant
by a girdle of silk. The butterflies
appear in spring and the male is
particularly noticeable as it flutters
in the sunshine. The female lacks
the orange spots of the male and
somewhat resembles one of the
white butterflies (e.g. No 53). Eggs
are laid singly on the flower stalks of
cuckoo-flower and other crucifers.
The caterpillars closely resemble
seed-pods. Cannibalism frequently
occurs amongst the early instars.

5 Green-veined White
Pieris napi
Ws 37–47 mm　April–Sept
Common; meadows, gardens, woods

This butterfly can be distinguished
from other whites by the dark-
powdered veins on the undersides of
the hind-wings. It hibernates as a
pupa and the adults emerge in the
spring. There are usually two generations. The green or bluish-green
caterpillar has black warts and light-
coloured lateral streaks. It feeds on
various crucifers e.g. cabbage,
charlock and swede causing some
damage. The pupa is usually green
or brownish and its colour varies with
the background. The adults fly even
on dull days.

6 Black-veined White
Aporia crataegi
Ws 56–67 mm　June–July　Open
places in wooded areas

Extinct in Britain but widely distributed on the Continent and occasionally
an orchard pest. The female lays eggs
on hawthorn, blackthorn, plum and
apple trees. The caterpillars hibernate gregariously inside a thick
cocoon fastened in a fork of the tree.
In spring they eat the new foliage
and, if disturbed, crawl quickly to
its densest parts. The caterpillar is
grey with a black back and orange-
yellow lateral stripes. When almost
fully grown, the caterpillars disperse,
attach themselves individually by
threads to the thicker branches, and
pupate.

7 Small Tortoiseshell
Aglais urticae
Ws 35–52 mm   Mar–Oct   Gardens
fields, meadows, glades.

One of our commonest and best
known butterflies. The adult hibernates in outhouses and can be seen on
the wing on sunny days in March. It
is a frequent garden visitor, attracted by plants with large amounts of
nectar (such as *Buddleia* – see plate).
The female lays 50–100 eggs on
the under-side of stinging nettle
leaves. The caterpillars live gregariously in silken tents like those of
the "peacock butterfly" (No 101).
They are dark with yellow lateral
bands and tubercles.

8 Comma
Polygonia c-album
Ws 40–50 mm ⋅ March–Aug
Common; gardens, woodland

This species resembles a dead leaf as
it hibernates on a tree trunk or in a
crack on a wall. Eggs are laid in the
spring on willows, hops, nettles,
currants and elm. The caterpillar is
red-brown with the posterior half of
the back white and is furnished with
orange-yellow spines. The pupa
hangs from a leaf upside down and
the adult emerges after about ten
days. The white C-shape on the
under-side of the hind-wing gives the
species its name.

9 Painted Lady
Cynthia cardui
Ws 51–60 mm   June–Sept   Fields,
meadows, gardens

Migratory. Large numbers of this
butterfly travel from their winter
quarters in N. Africa northwards
across Europe, and a few individuals
may even reach the shores of the
Arctic Ocean. Numbers reaching
Britain vary from year to year. Eggs
are laid on thistles, nettles and
burdock and the new generation
appears in late summer. It cannot
overwinter in Britain, however. There
is evidence that some return south
in autumn. This species has the
widest distribution of any butterfly
and is absent only from S. America.

10 Red Admiral
Vanessa atalanta
Ws 52–62 mm   May–Oct   Fields, meadows, gardens.

This species is migratory, and very rarely survives winter in Britain. Each year, in June, it arrives from S. Europe. The female lays eggs singly on upper-sides of nettle leaves and when the larva hatches it fastens the edges of the leaf together with silken threads. It feeds inside this shelter until it is destroyed and then makes a new one. The caterpillar is yellowish-green to dark brown and has light lateral spots and yellow spines. In autumn the new generation appears.

11 Peacock
Inachis io
Ws 52–63 mm   May–June
Common; fields, meadows, gardens

This species hibernates as an adult in buildings etc. If disturbed it makes a hissing noise by rubbing its wings together. It wakes early in the spring and the male chases the female in a complicated mating flight. It sometimes alights with its wings outspread to bask in the sunshine. The eye-like markings may frighten enemies especially as they appear and disappear rapidly as the butterfly opens and closes its wings. Eggs are laid in batches on the underside of nettle leaves. Caterpillar No 101.

12 Camberwell Beauty
Nymphalis antiopa
Ws 61–77 mm   Apr–May, Aug–Sept
Woodsides and clearings.

This species, in Britain, is a rare vagrant but is widely distributed on the Continent. It is attracted to sap exuding from trees and to rotting fruit. Eggs are laid in rings around twigs of sallow, osier willow, aspen or birch. The caterpillar is black with rust-red spots and black spines. It is voracious and gregarious; a colony can rapidly defoliate an entire branch. The pupa lasts for about three weeks and the adult lives for up to ten months hibernating in hollow trees etc.

13 Dark-green Fritillary
Argynnis aglaja
Ws 45–58 mm   July–Aug   Locally common; flowery slopes

The name fritillary comes from Latin and means dice-box, a reference to the resemblance of these insects to old checker-pattern inlaid dice-boxes. The upper-sides of the wings are reddish-brown with black markings. Fritillaries are especially attracted to purple flowers; this species lays its eggs on various species of wild violet. Hibernation occurs in the egg or early larval stages. The full-grown caterpillar reaches a length of 45 mm and is blackish-grey with red spots.

14 Lesser Marbled Fritillary
Brenthis ino
Ws 34–42 mm   June–Aug
Meadows and marshy places

This species does not occur in Britain although it inhabits much of central Europe. Its host plants are meadowsweet, raspberry, blackberry and cloudberry. The larvae hibernate at the base of the plant. They awake in spring and eat the new leaves. The pupa is yellowish-brown with pointed knobs and, as with other fritillaries, it is suspended from the food plant by its tail.

15 Small Pearl-bordered Fritillary
Boloria selene
Ws 35–41 mm   June–Sept
Common; woodland margins and clearings

The females of this species can often be seen flying just above the vegetation in search of suitable plants for oviposition. The caterpillars live mainly on violets. They are blackish-brown with a double light-coloured stripe and yellow spines and grow to 30 mm in length. They hibernate rolled up in dead leaves on the ground. In May they pupate and after about two weeks emerge as adults.

16 Heath Fritillary
Mellicta athalia
Ws 29–42 mm   June–July
Meadows

Common in continental Europe and S.
England this species lacks the glossy
spots on the under-side of the wings
that characterize the fritillaries of
other genera. The female lays eggs
in batches mainly on common cow-
wheat but also on other plants. The
gregarious caterpillars are dark with
white spots and are furnished with
brown white-pointed spines. They
grow for some time and then descend
to the ground to hibernate under a
silken web. In spring they reappear
and after feeding to maturity, pupate
on the food plant.

17 Scarce Copper ♂
Heodes virgaureae
Ws 26–32   July–Aug   Meadows,
glades

This species is common in central
and northern Europe but is absent
from Britain. The female differs from
the male in having the upper-side of
the wings paler with dark markings.
The eggs are laid on the ground.
These are in diapause and hatch in
spring, the larvae developing on
dock. The full-grown caterpillar is
green with short, red hairs, light-
coloured dorsal tubercles and lateral
stripes. The pupa is attached to the
stem of the host plant. If disturbed it
can make a faint hissing noise.

18 Small Copper
Lycaena phlaeas
Ws 22–32 mm   May–Oct   Common;
gardens, meadows, glades.

This lively little butterfly usually
stays within a strictly delimited area
and defends this territory against
intruders of the same species. The
sexes are similar and the eggs are
laid on plants of the dock family. The
colour of the caterpillar varies
with that of the leaves of the host
plant and, as they are similar in shape to outgrowths of the plants, are
well camouflaged. Two, or even
three, generations a year occur in
Britain.

19 Large Skipper
Ochlodes venata
Ws 26–32 mm   June–July
Common; meadows

Skippers differ from other butterflies in having a broad head. Their name is derived from their characteristic darting flight. The eggs of this species are laid on various grasses and reeds. The caterpillar lives within a tent of grass and winter is spent in this stage. When full-grown it spins a cocoon incorporating growing and bitten-off grass. The pupa has small hooks to hold itself in position within this cocoon. This species occurs throughout the temperate regions of Europe and Asia.

20 Dingy Skipper
Erynnis tages
Ws 25–28 mm   May–July   Woodsides and meadows.

This small inconspicuous species is common in woodlands. In spring it is commonly seen sunning itself on grass or sucking nectar from flowers. Resting on bare ground, it is particularly well concealed because of its colouring and markings. At dusk, or when it rains, the wings are folded over the body like a hood. The female lays eggs on birdsfoot trefoil, on which the larvae spin a shelter of several leaves. The full-grown caterpillar hibernates in this shelter which also houses the pupa.

21 Grizzled Skipper
Pyrgus malvae
Ws 20–24 mm   May–June
Common; meadows, hillsides

This is such a fast flier that it is difficult to follow it in flight. It rests with its wings outstretched, frequently on the ground. The female lays eggs on bramble, wild raspberry, wild strawberry and cinquefoils. After hatching the caterpillar spins a shelter over itself coming out only after dark. Later it spins leaves together to form a small tube. When full-grown it crawls into a rolled leaf at the base of the plant to pupate. This is the only skipper that overwinters as a pupa.

22 Common Blue
Polyommatus icarus
Ws 21–33 mm   June–Aug   Fields and downs

This species is a common sight on sunny summer days as it flies from flower to flower in search of nectar. Occasionally mating can be observed (see plate). The caterpillar feeds on birdsfoot trefoil, clover etc. and consumes the whole plant, leaves, flowers and pods. Winter is spent as a half-grown caterpillar and pupation occurs in spring after a further bout of feeding. There may be one or two broods according to latitude.

23 Mazarine Blue ♂
Cyaniris semiargus
Ws 24–33 mm   June–Aug
Meadows, mountain slopes

This species no longer occurs in Britain but is common over much of the continent. The females differ from the males in having the upper-side of the wings brown instead of blue. The caterpillar lives on clover and kidney vetch. It hibernates in a dead flower in which it remains for about seven months. It resumes feeding on the young shoots in the following spring. Pupation takes place just above the ground and lasts about three weeks.

24 Silver-studded Blue ♂
Plebejus argus
Ws 23–27 mm   June–Aug.
Common; meadows and heaths

Blues are most easily identified by the markings on the undersides of the wings. The caterpillar of this species lives on heather, broom and gorse. It is greenish, and like that of most other blues, is short and stubby like a woodlouse. The caterpillars of all blues secrete a sweet fluid that attracts ants. In one species, the "large blue" *Maculinea arion*, the caterpillars are myrme-cophilous, i.e. they live in ants' nests, feeding on the eggs and young of the ants.

25 Six-spot Burnet
Zygaena filipendulae
Ws 28–39 mm   May–Aug   Common;
meadows, downs, marshes

Burnets are slow-flying moths and
groups of them may often be seen
resting on flowers. Their bright
colour is apparently a warning to
potential predators that they are
distasteful and birds leave them
alone. Both adults and larvae secrete
a noisome, poisonous yellow fluid
and their tissues contain hydrocyanic
acid. The eggs are laid on legumes,
the caterpillar is yellow-green and
hibernates at an early stage. The
yellowish cocoon containing the pupa
is usually attached to a stalk of
grass.

26 Forester
Procris statices
Ws 24–32 mm   May–June   Local;
damp meadows

This moth can be seen in summer
visiting flowers such as ragged robin
and scabious. It belongs to the burnet
family whose members resemble butterflies in having clubbed antennae.
Eggs are laid on docks and the caterpillars first mine the leaves but later
feed on their under-sides so that the
upper surface remains intact. The
caterpillar is hairy and is greyish-
yellow with a brown dorsal stripe and
broad reddish lateral ones. After
hibernation it pupates in a white
cocoon either on the plant or on the
ground.

**27 Broad-bordered Bee Hawk-
      moth**
Hemaris fuciformis
Ws 37–48 mm   May–June   Common
locally; meadows, gardens

This species differs from the large
hawk-moths in being active only by
day. It flies to rhododendrons, lilac,
bugle and other flowers and hovers
before them sucking nectar by means
of its long proboscis. It somewhat resembles a bumblebee – an example
of protective mimicry. Eggs are laid
on the underside of leaves of honeysuckle. The caterpillar is green with
red markings and has a 'horn' on its
tail. It closely resembles the food
plant in appearance. The pupa lies
on the ground in an earth-covered
cocoon.

28 Small Elephant Hawk-moth
Deilephila porcellus
Ws 43–52 mm   May–July
Common; fields, meadows, gardens

This species starts flying shortly after sunset and visits various flowers. Mating takes place later in the evening. The female sits on vegetation and the male flies round her, touches her with his antennae and flies off. This is repeated several times and then copulation takes place suddenly. The female lays about 100 eggs on various plants e.g. bedstraws and willow-herbs. The caterpillar is dark-brown with large eye-like spots behind its head. It lacks, however, an anal 'horn'.

29 Silver Y
Autographa gamma
Ws 33–41 mm   June–Sept
Meadows, fields, gardens

This belongs to the Noctuidae. Unlike most members of this family it flies by day as well as by night. It probably does not survive winter in Britain but large numbers migrate from the Continent during spring and summer each year. Each female may lay 500 eggs and the caterpillars eat the foliage of many kinds of plants. When numbers are high damage may be inflicted on crops, in particular, on sugar-beet. Pupation takes place in cocoons on foliage and a new generation appears in August.

30 Antler
Cerapteryx graminis
Ws 25–37 mm   July–Aug   Common; meadows, fields

This species flies by day as well as by night and may be seen resting on flowers. The female, considerably larger than the male, lays eggs while flying over grassland, usually on rough hill pasture. The greyish-brown or dark-brown caterpillars attack the bases of plants causing them to die and, particularly in northern parts of the country, they may become so numerous that large areas of grassland are laid bare. The pupa is found in the soil and winter is spent in the egg stage.

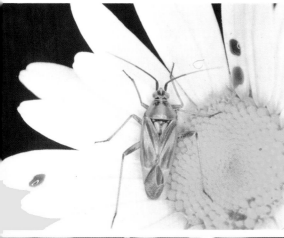

31 Capsid bug
Calocoris roseomaculatus
L 6–9 mm   June–Oct   Common;
on flowers, especially composites
and legumes

The entomologist reserves the use of
the word ''bug'' to mean a member
of the group to which the capsids
and their relatives belong. These insects all possess mouth-parts which
are adapted for piercing and sucking.
This species feeds on plant juices
and pollen grains which it first
mixes with saliva. Some species,
such as the ''potato capsid,''
Calocoris norvegicus, can cause
appreciable damage to crops. The
eggs are laid in autumn in crevices
on buds or bark where they hibernate. There is no pupal stage.

32 Longhorn beetle
Strangalia maculata
L 13–18 mm   July–Aug   Common;
on flowers e.g. Meadowsweet and
umbels

This and the following beetle are
nectar and pollen feeders commonly
met with on flowers. The name of the
family, the longhorn beetles refers
to the sometimes very long antennae
that characterize this group. The
larvae live in tree stumps and rotting
branches of various deciduous trees
e.g. birch. This species becomes rarer
towards the north, but occurs
throughout the British Isles. There
are rather more than 60 species of
longhorns found in Britain.

33 Longhorn beetle
Leptura rubra
L 12–18 mm   July–Aug   Local; on
flowers in woodland

The larvae of this species occur in fir
stumps and attack wooden fences at
a few localities in Britain. The
differences between the sexes are
striking. The male (above) is small
and has the elytra reddish-yellow
while the female has them red.
There are six species of this genus in
Britain none of which are really common.

34 Soldier-beetle
Cantharis livida var. *rufipes*
L 11–15 mm   May–June   Common;
on flowers in woods and meadows

There are about 16 soldier-beetles
of the genus *Cantharis* in Britain.
They are predatory as adults on various small insects and may be seen
flying slowly or resting on flowers,
particularly umbels. The species
shown here occurs in two colour
variants in Britain. One is as in the
plate and the other, more common,
is all brown. The larvae dwell in
the soil and have a velvet-covered
appearance (No 240).

35 Chafer-beetle
Trichius fasciatus
L 10–16 mm   June–July   Rare; on
flowers, mainly umbels

This beetle closely resembles a
bumblebee in appearance – an
example of protective mimicry.
Although common in parts of the
Continent e.g. over much of Scandinavia, the species is rare in Britain
and appears confined to a few localities in Scotland and Wales. The
larvae live in decaying wood and may
be found in the dead stumps of deciduous species such as birch and
alder.

36 Chafer-beetle
Cetonia cuprea
L 17–22 mm   June–Aug   Local;
in Scotland and N. England on
flowers

This chafer flies only in bright sunshine. Unlike most beetles, it can
fly without raising its elytra as these
are cut away at the sides to allow
the hind-wings to be protruded. The
adults eat petals, the larvae occur
in the nests of ants, mainly of *Formica rufa* the "wood-ant". The ants do
not usually attack them, but if they
do, they secrete a noxious fluid. A
close relative, the "rose-chafer",
C. aurata occurs in S. England where
its larvae sometimes damage pasture.

37 Wasp-beetle
Clytus arietis
L 10–14 mm   June–July   Common;
on flowers and trees

This beetle is common throughout the British Isles and much of Europe. It is a good example of protective mimicry as it resembles a wasp both in appearance and activity. This situation, in which a harmless animal imitates a poisonous or distasteful one is known as "Batesian mimicry" after the 18th C. British naturalist H.W. Bates who described it. *C. arietis* is a swift and agile flyer attracted by warmth and sunlight. The larvae live in dead branches and cause damage by boring into and weakening fence posts, etc.

38 Common wasp
Vespula vulgaris
L 11–20 mm   April–Oct   Common;
on flowers and fruit

The fertilized female comes out of hibernation in the spring and searches for a suitable nest site. A new colony is established in a hole in the ground, a hollow tree, or perhaps under the eaves of a house. The colony consists of the queen, a number of males, and an increasing number of workers (sterile females). The total population of a wasp nest may exceed 10,000. In warm and dry summers, wasps can become a real nuisance.

39 Hornet
Vespa crabro
L 20–35 mm   April–Oct   Locally
common; in S. England

The "hornet" is our largest species of wasp. It rarely if ever stings except in defence, but several stings can be fatal, especially for children or allergic adults. The "hornet" gnaws dry wood and mixes it with saliva to produce a grey paper-like material from which it constructs its nest. This has an opening in the bottom and consists of a large number of hexagonal cells in each of which the female lays a single egg. The nests are often built in hollow trees.

40 Bumblebee
Bombus lucorum
L 10–22 mm   April–Oct   Common;
on flowers

The bumblebee hibernates as a fertilised female. In spring she builds a nest in the ground, often using an old mouse hole. After lining this with moss etc. she makes a waxen cup for pollen and eggs and a second one for honey. The larvae feed on these supplies which are continually renewed. The first brood consists of sterile females – workers – that share the task of feeding the colony. In autumn males and fertile females are produced. After pairing the males die and the females go into hibernation.

41 Bumblebee
Bombus agrorum
L 13–19 mm   April–Oct   Common;
on flowers

Bumblebee colonies contain 100–100 individuals when fully developed. These insects are important as pollinators of fruit trees and field crops and poor yields of clover and lucerne seed have been attributed to their populations having been reduced by agricultural operations such as ditch filling and scrub clearance which tend to reduce nesting sites. Insecticides also take their toll. The species illustrated is one of the commonest of the nineteen found in Britain.

42 Honeybee
Apis mellifera
L 15–21 mm   March–Nov   Common;
on flowers

The "honeybee" has been domesticated since antiquity. A large colony can consist of a queen, about 200 males and some 70,000 workers. It is a miracle of organization and division of labour. The workers build combs, hexagonal cells of wax secreted by glands on the under-side of the abdomen. They produce honey frcm nectar they gather from flowers and collect pollen to feed their larvae. When a new queen is born the old one leaves the hive with a large number of workers and starts a new colony.

43 Slender-bodied Digger-wasp ♂
Crabro cribrarius
L 10–16 mm June–July
Common; flowers in sandy areas

Digger-wasps have silvery or golden pubescence on the face. The males of this and closely-related species have a peculiar shield-like appendage on the fore-legs. They derive their name from their habit of digging their nests in sandy ground. They live on flies that they catch in flight. Some of these are stored in the nests as food for the larvae.

44 Crab-spider ♀
Misumena vatia
L 3–10 mm April–Aug Common; on flowers

This spider sits in flowers waiting for insects to alight. When one does (plate) the spider attacks and grasps it with its fore-legs and proceeds to suck its body juices. The species is well concealed because its body can assume the colour of the flower on which it sits, taking about two days to change completely from yellow to white or vice versa. The male, however, is always green. This spider's eggs are laid in a sheath consisting of two leaves spun together.

45 Bee-fly
Systoechus sulphureus
L 5–8 mm On flowers

Bee-flies are good flyers and can remain absolutely still in mid-air while they suck nectar from flowers. The larvae are parasites on bees and solitary wasps. Some are hyperparasites, i.e. they develop inside the bodies of other parasites. The species shown here is a continental form not found in Britain, but several of its close relatives are. The most common of these is the large *Bombylius major* which is active from mid-March until early May in the southern half of the country.

46 Hover-fly
Volucella bombylans
L 10–22 mm   May–Aug   Common;
on flowers

This species lives on nectar and pollen from various flowers. It closely resembles a bumblebee but is a true fly bearing only one pair of functional wings. This is, of course, a case of mimicry, the model being *Bombus lucorum* (No 40) which has a powerful sting. Different individuals mimic different species of bumblebees. The larvae are predators of bumblebee grubs and presumably the different colour varieties are connected with the species the larvae prey on.

47 Hover-fly
Volucella pellucens
L 11–16 mm   May–Sept   Common;
on flowers

This powerfully built fly is often seen on composites and umbels. It lays its eggs in wasps' nests and the larvae live on dead wasps and the left-overs of the wasps' food. They serve, in fact, as house cleaners in the wasp nest. When full-grown, they leave the nest and pupate in the soil.

48 Hover-fly
Rhingia campestris
L 8–11 mm   April–Nov   Common;
on flowers

There are over 230 species of hover-fly in Britain. They are often beautifully coloured and differ greatly in body shape and size. The food of the adults generally consists of nectar and pollen. This species has a long proboscis used for obtaining nectar from deep flowers, such as hair bells and columbine (plate). Eggs are laid on vegetation around cowpats, etc. The larvae hatch and drop onto the excrement and burrow in it. They grow rapidly and emerge in large numbers towards the end of summer.

49 Hover-fly
Syrphus balteatus
L 7–11 mm Feb–Nov Common; on flowers

This sun-loving species is often abundant in gardens. It belongs to a group of hover-flies whose larvae are predacious on aphids. The maggot-like larvae crawl over the plants in search of aphids and, piercing them with their mouthparts, suck out their juices. One larva can kill up to 100 aphids in a day. When fully grown the larva pupates within the inflated and hardened last larval skin which is cemented to a leaf.

50 Drone-fly
Eristalis tenax
L 11–16 mm March–Nov Common; on flowers

This fly has an almost world-wide distribution. It closely resembles a "honeybee" and sounds like one when it flies. It is very common, especially in autumn, and often enters houses. The female lays eggs near manure heaps or near stagnant pools and the larvae live in the nutrient-rich effluent. As this environment is deficient in oxygen the larvae possess a long "snorkel"-like breathing tube which enables them to breathe air directly. These larvae are known as "rat-tailed maggots".

51 Green bottle
Lucilia species
L 5–11 mm April–Sept Common; on flowers, meat and excrement

Green bottles and the related blue bottles have, like hover-flies, one pair of functional wings. The second pair have been modified to form club-shaped balancing organs, the halteres. The commonest species in Britain is *Lucilia sericata* which is the main cause of "primary strike" in sheep. The flies lay eggs in soiled parts of the fleece and the larvae (maggots) eat into the flesh causing extensive wounds. These attract other flies, including blue bottles, and their maggots extend the damage. Severely afflicted sheep may die if untreated.

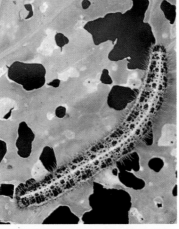

52 Large White (eggs and caterpillar)
Pieris brassicae
Caterpillar L 35–45 mm   April–Sept
Common; on crucifers and garden nasturtium

This butterfly lays eggs in batches on leaves of crucifers and garden nasturtium. The caterpillar hatches in about a week and starts to eat the leaves. At first it bites isolated small holes but gradually the whole leaf becomes consumed. Great damage, particularly to crops in gardens, may be caused. After about a month the caterpillar seeks a vertical surface on which to pupate. It attaches itself by means of a silken girdle and tail pad and pupates head uppermost. Caterpillars and pupae may be parasitised by minute wasps.

53 Large White ♀
Pieris brassicae
Ws 50–64 mm   April–Sept
Common; in gardens, fields etc.

This species hibernates as a pupa. There are two or three broods a year, the first often laying eggs on wild crucifers while the second, emerging in July, tends to lay on cultivated plants. British populations are augmented by migrants from the Continent and where numbers are high damage to crops is greatest. Coastal regions are particularly susceptible.

54 Small White ♀
Pieris rapae
Ws 42–47 mm   Feb–Sept   Common; in gardens, fields etc.

This species is distinguishable from the "large white" (No 53) by its smaller size and by the much smaller black patches on the tips of the fore-wings. The male differs from the female in having only one dark spot on the fore-wing. In some areas it is commoner than the large white. It has a similar life history and can also be a serious pest of crucifers. Unlike the large white, the eggs are deposited singly and the solitary caterpillars are all green.

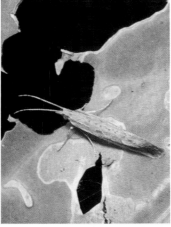

55 Diamond-back
Plutella xylostella
Ws 13–15 mm   May–Aug   Common;
on crucifers

The caterpillars of this moth can
cause severe damage to cultivated
crucifers such as cabbage and turnip.
The young ones mine the leaves and
older ones feed on the under-sides
so that only the upper skin is left
intact. Pupation is in a silken cocoon
fixed to the under-side of a leaf.
Winter is spent as an adult and there
are 2–3 generations a year in Britain,
the caterpillars being evident from
May–Aug. Numbers in Britain may
at times be augmented by immigrants
from the Continent.

56 Cabbage Root-fly
Erioischia brassicae
L 5–8 mm   April–Sept   Common; on
crucifers

There are two very similar species,
the small *E. brassicae* and the somewhat larger *E. floralis,* both economi-
cally important pests. They lay eggs
in the ground close to cabbages,
turnips etc. When these hatch, the
larvae attack and burrow into the
plant roots. The plants become
yellow and wither. The plate on the
left shows larvae in a swede. *E.
brassicae* has 2–3 generations in a
season, *E. floralis* has only one. They
both hibernate as pupae.

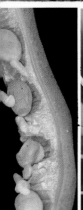

57 Brassica Pod-midge
Dasyneura brassicae
L 1–2 mm   May–Sept   Common; on
crucifer seed-crops

The larvae live in seed-pods of
crucifers such as mustard grown for
seed. The females cannot pierce the
pods so they use the holes made by
the "brassica seed-weevil" (No 59),
when laying their eggs. The infested
pod swells, becomes deformed and
dies (right). The larvae (left) eat
the seeds and walls of the pod.
There are 3–4 generations a year
and winter is spent as a mature larva
in the soil. An unusual feature is
that larvae in the pods reproduce
without becoming adult.

58 Blossom-beetle
Meligethes aeneus
L 2–3 mm   April–Sept   Common; on yellow flowers

Wherever crucifers are grown for seed this beetle is likely to become a pest. On the Continent it is a pest of oil-seed crops, in Britain it is important where cabbages etc. are grown for seed. The adult hibernates. In spring it feeds on pollen of wild flowers and then flies to crucifers where it eats flower-buds and lays eggs inside them. The larvae feed on pollen and after 2–4 weeks fall to the ground and pupate. In late summer new beetles emerge and after feeding, hibernate.

59 Brassica Seed-weevil
Ceuthorhynchus assimilis
L 2–3 mm   April–Sept   Common; on crucifers

Often occurs together with "blossom-beetle" (No 58), but is more difficult to observe. If disturbed it releases its hold and drops to the ground where it remains immobile for some time. It hibernates as an adult. In the spring it feeds on soft plant tissues and lays eggs in the newly formed seed-pods. The larvae feed on the seeds causing considerable damage. The holes made in the walls of the pods facilitate the entry of the "brassica pod-midge" (No 57).

60 Small Striped Flea beetle
Phyllotreta undulata
L 2–3 mm   April–Sept   Common; mainly on crucifers

There are 16 representatives of this genus in the British Isles. Several species are black and yellow, as in the plate, others are black or green, often with a metallic sheen. Some are pests of crucifers and may transmit plant virus diseases. Others occur on cereals. They hibernate as adults under stones, bark etc. In spring they attack seedlings, such as cabbage and radish making small holes all over the leaves. If disturbed, the beetles jump like fleas. The larvae feed on the roots of the food-plants.

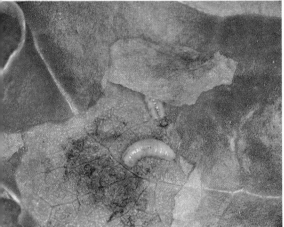

61 Beet Leaf-miner (larvae)
Pegomya hyoscyami
L 7–8 mm   April–Oct   Common; on
beet, mangold and spinach

It lays eggs on the under-side of beet
leaves. The larvae penetrate the leaves and eat the tissues forming a
"mine". In the plate the leaf surface
has been peeled away to expose the
larvae. After about two weeks the
larva is full-grown and drops to the
ground to pupate. In Britain there
are three generations in a summer.
The first causes the greatest damage
as the plants at this time are small.
The pupae of the last generation
hibernate in the soil.

62 Black Bean-aphid
Aphis fabae
L 1–3 mm   March–Nov   Common;
on various plants

This aphid hibernates as eggs on
spindle tree, mock orange and guel-
der rose. From these hatch wingless
"stem-mothers". These are all
female and give rise to several
generations of females without
fertilization (parthenogenesis). Soon
large numbers build up and winged
females are produced which fly to
summer hosts, e.g. broad bean and
beet. They weaken the plants by
removing large amounts of sap and
by introducing virus diseases. Aphids
have many enemies, e.g. ladybirds
and hover-fly larvae, but their
reproductive potential is enormous.

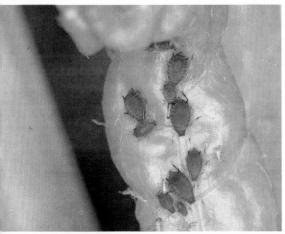

63 Grain-aphid
Macrosiphum avenae
L 2–3 mm   March–Sept   Common;
on grasses and cereals

This species attacks the leaves and
later the developing grains in the
ears of various cereals. It also transmits virus disease. The fertilised
female lays eggs on various grasses
in autumn and these hatch in March
to produce stem mothers. After several parthenogenetic generations
winged forms appear in June and
fly to cereal crops. In autumn the
sexes appear and once more eggs
are laid on grasses. The paired
processes on the abdomen, the
cornicles, secrete a liquid containing
an alarm pheromone when the aphid
is attacked.

64 Colorado-beetle
Leptinotarsa decemlineata
L 9–11 mm   May–Sept   On potato

This beetle came to Europe from North America in 1922. It now occurs in almost the whole of continental Europe and in recent years has spread into Scandinavia. So far it has not gained a foothold in Britain thanks to rigorous quarantine regulations. The adult beetle hibernates deep in the ground, a situation which might limit its further spread northwards. The female lays 500–800 eggs and the red larvae (left), which are voracious feeders, can soon lay waste entire potato fields. The adults also feed on potato leaves.

65 Rosy Rustic (caterpillar)
Hydraecia micacea
L 35–45 mm   May–Aug. Common; mainly on potatoes

The eggs are laid in the lowest blades of grass in August–September and hatch in the following spring. The larvae live inside the stems of various plants, such as beet, rhubarb, iris and potato. The infested plants usually wither. Pupation takes place in late summer and the adult moths appear in late August. The colour of the moths varies from greyish- to reddish-brown. They are usually found in cultivated areas and near swamps. The female is larger than the male.

66 Potato Cyst-nematode
Heterodera rostochiensis
L 0.5–1.5 mm   All year   Common on roots of potato and tomato

Nematodes are white, unsegmented worms but in this species the female swells to form a brown, spherical, egg-filled cyst (plate). When the eggs hatch, the larvae penetrate the fine roots of the plant and grow within them. On maturation the worm-shaped males escape but the females remain attached to the roots. Infested plants are stunted with rolled leaves and produce little crop. The nematode can survive in the soil for several years and eradication is difficult. Nowadays, potatoes can be produced that are resistant to attack.

67 Willow-carrot aphid
Cavariella aegopodii
L 2–3 mm   March–Oct   Local; on sallow, willow, carrots, parsnips, etc.

The winter hosts of this species are sallow and osier willows from which it migrates in spring to umbelliferous summer hosts such as the carrot. Like most other aphids it builds up enormous numbers in warm dry weather. Because of substances the aphids inject into the plants via their piercing mouth-parts, the leaves become curly and deformed and the plants may become stunted or killed.
This aphid is also the vector of virus diseases that cause discolouration and stunting of the plants.

68 Carrot-fly (larva)
Psila rosae
L 6–8 mm   May–April   Common; on carrot

The larvae of "carrot-fly" are the "worms" of carrot. The fly lays eggs at the base of small plants. The larvae eat out cavities in the carrot, the leaves become yellow and the plant often dies. The larvae become full-grown after about a month then crawl an inch or so into the ground and pupate. A second generation is active in September and its progeny pass the winter as larvae or pupae. A simple and fairly effective counter measure is to plant the carrots relatively late.

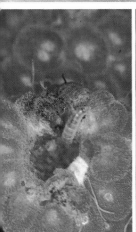

69 Raspberry-beetle
Byturus tomentosus
L 3–5 mm   April–July   Common; on raspberry and related plants

We have all seen the larvae of this beetle (left) when picking or eating raspberries. The adult (right) a little brown beetle, lays its eggs in the spring in the raspberry flowers and the larvae feed on, and spoil the fruit. When mature they crawl into the soil and pupate. Cultivated raspberries can be protected by spraying the flower buds with a preparation that is harmless to beneficial insects such as honey- and bumblebees.

70 Magpie-moth
Abraxas grossulariata
Ws 35–41 mm   July–Aug   Common;
gardens, woodsides

The female lays eggs on the under-
sides of leaves of gooseberry or
currant bushes and on blackthorn,
hawthorn and plum. The eggs hatch
after a few weeks and the young
caterpillars begin eating the leaves.
They hibernate between leaves that
have been spun together on the
ground and continue feeding the
following spring. They pupate in
early summer and the moths appear
in July. They vary in colour from
almost completely black to almost
completely white. The caterpillars
are beautifully marked with black,
white, and yellow spots.

71 Gooseberry sawfly (larva)
Nematus ribesii
L 15–20 mm   April–Oct   Common;
on gooseberry and currant bushes

The yellow and black sawfly lays
its eggs in a string along the midrib
on the under-side of leaves as soon
as these come out in spring. After
about ten days the larvae appear
and within a short time can defoliate
whole bushes, causing the half-ripe
fruit to be shed prematurely. There
can be 2–3 generations in a summer.
The larvae hibernate inside a spun
cocoon in the soil. They pupate in
the following spring and the adults
usually emerge in April or May.

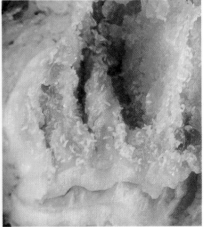

72 Black Currant Gall-mite
Cecidophyopsis ribis
L about 0.2 mm   All year   Common;
on black currants

In spring one comes across black
currant bushes with swollen buds.
These do not form shoots but wither
away. On the left are a normal bud
and an attacked one (gall). Each gall
contains up to 30,000 minute white
4-legged mites (right-dissected gall).
Later in spring the nymphs leave the
gall and are dispersed by wind or
pollinating insects. They find new
buds in which they multiply rapidly.
The mites carry the virus disease
'reversion' which renders the bushes
unthrifty. Substantially infested
bushes are best burned.

73 Large Rose sawfly
Arge ochropus
L 7–10 mm June–Aug Common; on roses in S. England

Occurs on both wild and cultivated roses. The eggs are laid in cracks in buds or petioles in rows of about 15–20 at a time. The shoots on which the eggs are laid soon become brown and deformed. The larvae (left) feed on the leaves and can cause serious damage. They pupate in the soil and there may be two generations in one season. The adult is often seen on rose bushes sitting flat against the surface of a leaf apparently to catch warmth from the sun.

74 Rose-aphid
Macrosiphum rosae
L 2–3 mm May–Oct Common; on roses

Extremely common on cultivated roses. They form dense colonies on young shoots and buds particularly in the early summer. They are either green or reddish. The females that hatch in the spring lay young (see plate) without being fertilized (parthenogenesis) and so do the succeeding generations. When over-crowding occurs, winged females are produced which fly off in search of new rose bushes. In the autumn males and egg-laying females are pro-duced. There is no alternation between winter and summer host plants.

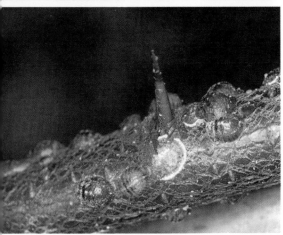

75 Brown-scale
Parthenolecanium corni
L 3–5 mm All year Common; on trees and bushes

The females look like small brown outgrowths attached to bark of trees or bushes. They have enormously elongated mouth-parts with which they penetrate the vascular tissues of the host plant and gain access to a rich and continuous supply of food. This species, like most of its relati-ves, is parthenogenetic (see No 74). The eggs are concealed under the shield of the mother insect although there can be as many as 3,000. Hibernation takes place in the larval stage. Usually there is little indica-tion that the plant is damaged.

76 Mealy Plum-aphid
Hyalopterus pruni
L 2 mm May–Sept Common; on plum, blackthorn, aquatic grasses and reeds

Most summers large numbers of this aphid can be found on the undersides of plum leaves. The leaves become sticky and discoloured as a result of the copious amounts of liquid excretion (honeydew) that this species, like many of its relatives, produces. About midsummer winged forms appear and these fly to the summer hosts, aquatic grasses and reeds (plate) and in autumn some of their winged descendants return to the plum, produce males and females and lay over-wintering eggs on the shoots.

77 Apple sucker
Psylla mali
L 2–3 mm May–Nov Common; on apple trees

Apple suckers attack the opening flowers of the apple and destroy them completely thus spoiling the crop. In the past this damage has been erroneously attributed to frost. The young suckers are yellowish-green and flattened. They produce waxy threads. The adults (plate) are winged and jump and fly readily. Straw-coloured eggs are laid in cracks in the bark during August. These hatch in spring.

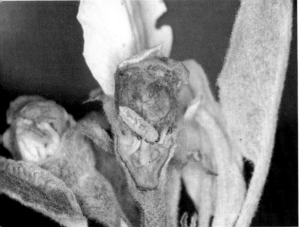

78 Apple Blossom-weevil (larva)
Anthonomus pomorum
L 3–5 mm May–July Common; on apple and pear trees

If a flower bud turns brown and dies it probably contains a larva of this weevil. The adult is about 5 mm long and greyish-brown. In spring it gnaws into a flower bud within which it lays an egg. This hatches and the larva feeds on the pistils and stamens of the unopened flowers. After about one month a new generation of adults is produced. These hibernate in bark cracks. However, when too many blossoms are produced, moderate attacks can have a beneficial thinning effect.

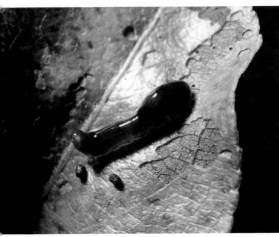

79 Pear and Cherry Slug-worm
(larva)
Caliroa cerasi
L 8–14 mm   June–Sept   Common;
on pears, cherries and whitebeam

These larvae are very distinctive.
They are really greenish-yellow but
cover themselves with a black, inky
slime which gives them the appearance of slugs. They are, in fact,
the larvae of a species of sawfly.
They live on the upper-sides of
leaves which they eat in such a way
that only the veins and the lower
cuticle remain. The adult is about
5 mm long and shiny black. It
appears in June after hibernating. It
is cosmopolitan.

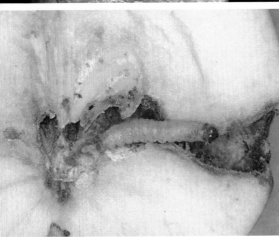

80 Codling-moth (caterpillar)
Cydia pomonella
L 20–25 mm   July–Nov   Common;
in apples, pears and walnuts

"Worm-eaten" apples may have
been attacked by these caterpillars.
Outside there is a hole containing
excrement, inside, a tunnel to the
core. If attacked early, the apple
withers and falls. If later, it may
mature but be unappetizing and
prone to rot. The larvae become full-
grown in one month. They hibernate
in crevices e.g. cracks in bark. The
adult emerges in summer. It is about
10 mm long, grey and brown with a
reddish spot on either fore-wing. It
lays eggs on fruits, leaves and stems.

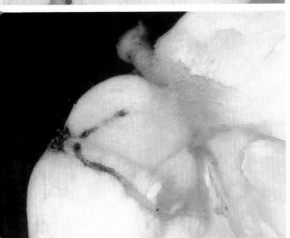

81 Apple Fruit-moth (caterpillar)
Argyresthia conjugella
L 7–8 mm   July–Sept   Common; in
rowan-berries and apples

The natural host of this species is
rowan, but apples are attacked too.
The caterpillars make narrow winding tunnels in the flesh of the fruit
the entrances of which appear as
small closed black spots on the skin.
The larvae live for a little over two
months and then crawl into the
ground to hibernate. Some pupate
before winter. The moths emerge in
May–June and lay eggs on the small
apples. They are about 10 mm long
and are yellowish-white with brown
markings.

82 Speckled Wood
Pararge aegeria
Ws 33–44 mm   May–Nov   Common;
shady places and woods

The family to which this butterfly belongs contains over 2,000 species throughout the world and some 113 in Europe. Most have round 'eye-spots' on the wings. They have a relatively slow, fluttering flight. The eggs are usually laid on blades of grass and the caterpillars feeding on these are difficult to find because of their green and brownish-yellow colours and nocturnal activity. Most species hibernate as caterpillars. The "speckled wood" is most often seen in shady clearings. In Britain there may be three generations.

83 Arran Brown
Erebia ligea
Ws 40–48 mm   July–Aug
Woodland

This species does not occur in Britain although there is an old record of it from Arran. Elsewhere in N. Europe it is a common woodland species. It hibernates as a caterpillar, the yellow-brown pupa lies directly on the ground in early summer and the adult appears in late summer. Like many butterflies, it is divided into several subspecies. That illustrated here, *Erebia ligea dovrensis,* lives in the birch region of the Scandinavian mountains.

84 Ringlet
Aphantopus hyperantus
Ws 32–45 mm   July–Aug   Common;
meadows and near woods

The "ringlet" will fly in dull weather although, like most other butterflies, it is most active in sunny weather. The upper sides of the wings are almost uniformly dark brown. The female drops her eggs on the vegetation while in flight and the caterpillars, which are mainly nocturnal, feed on various grasses. They hibernate at the roots, resume feeding in spring and eventually pupate in June. The adult emerges about two weeks later.

85 Grayling
Hipparchia semele
Ws 44–55 mm   July–Aug   Common; on dry, open grassland

A very shy and fast-flying butterfly. In flight it throws itself this way and that, and alights equally suddenly. When resting it becomes almost invisible since the prominent eye spots are covered by the mottled-brown hind-wings. The caterpillar is unique among British butterflies in that it burrows into the ground and pupates there in a cocoon. The photo shows a somewhat smaller and lighter subspecies, *Hipparchia semele tristis* which occurs on some of the Baltic Islands.

86 Meadow Brown
Maniola jurtina
Ws 36–49 mm   June–Sept   Road-sides, meadows, glades

One of the commonest butterflies but insignificantly marked and therefore often overlooked. It visits flowers, particularly blackberries, in search of nectar. On the fore-wings of the male are scales, the scent of which is even perceptible to humans. It is supposed to smell like the wood of a cigarbox. The caterpillars are light green with a dark stripe down the back and lighter stripes on the sides. Pupation takes place in spring the pupa being suspended by silken threads from a stalk of grass.

87 Small Heath
Coenonympha pamphilus
Ws 24–33 mm   May–Oct   Hillsides and fields

This small species is exceedingly common and may often be seen flying just above the vegetation, possibly in the company of blues and coppers. Towards evening, or when the weather turns dull it sits quite openly on the grass. Eggs are laid on grasses in May–June and some individuals develop quickly to produce a second generation in August. The rest develop more slowly. Caterpillars from both generations may thus hibernate and pupate together in the spring.

88 White Ermine
Spilosoma lubricipeda
Ws 34–45 mm   May–June
Common; fields and gardens

This moth, like many other nocturnally active species, is attracted to light and often gathers round street lamps and comes to lighted windows. The spots vary in number and form. The upper-side of the abdomen is yellow with black marks. In daytime it rests in low vegetation. The eggs are laid in compact batches on nettles and numerous other herbs. The caterpillars live on these plants from July–Sept and then pupate. The dark brown pupa overwinters on the ground within a cocoon of silk and hair.

89 Ghost Swift-moth ♀
Hepialus humuli
Ws 42–75 mm   June–Aug
Common; fields and gardens

A primitive moth with very short antennae, this species lacks a proboscis and is unable to feed as an adult. The sexes are very different in appearance. The male has shiny white wings and is considerably smaller than the female. The male flies at sunset over vegetation with a characteristic undulatory motion that attracts the female. After mating, the female drops her eggs from flight. The caterpillars feed on roots of many plants and may damage agricultural crops. They may take two years to reach maturity.

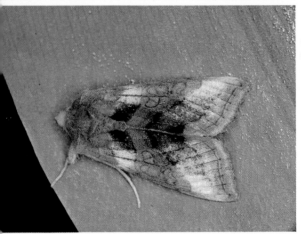

90 Burnished Brass
Diachrysia chrysitis
Ws 31–39 mm   June–Sept
Common; fields, gardens, woodsides

This beautiful shimmering moth sometimes visits flowers by day gathering nectar and may be seen flying along ditches and hedgerows when the flowers are in bloom. It is, however, attracted to light. The caterpillar is green with white markings and, having only three pairs of abdominal prolegs, crawls with a 'looping' motion (see No 93). It feeds mainly on nettles, hibernates at an early stage and resumes feeding in spring. Pupation takes place in May–June in woven cocoons attached to the host plants.

91 Heart and Dart
Agrotis exclamationis
Ws 30–40 mm   June–Oct   Common;
fields, gardens

The name of this common species is
derived from the markings on the
fore-wing. The caterpillars feed on
a wide range of plants and often bite
off seedlings of crop plants at ground
level causing damage of economic
importance. Farmers call the caterpillars of this, and of other species
with similar habits, "cutworms".
Winter is spent as a caterpillar and
pupation is in early spring in the soil.
There may be a partial second generation.

92 Common Heath ♂
Ematurga atomaria
Ws 20–31 mm   May–Aug   Common;
moors, heaths, marshes

This geometrid flies in broad day-
light and is abundant on heathland.
The female is slightly smaller than
the male; it is mottled grey instead
of brown and lacks the plumose
antennae. The caterpillars live mainly on heather. They vary in colour
from greenish-brown to grey with
lighter spots on the back. They have
only two pairs of abdominal prolegs
and progress with a looping motion
(see No 93).

93 Shaded Broad-bar
Scotopteryx chenopodiata
Ws 25–33 mm   July–Aug   Common;
fields, meadows, glades

This moth may often be flushed from
grass as one walks in a meadow but
soon settles and conceals itself in
the vegetation. The caterpillar lives
on various leguminous plants and is
greyish or brownish with black spots.
Like those of other species of geometrid, it has a characteristic way
of progressing by "looping". First
the tail end is brought forward so
that the body forms a hoop, then the
head end moves forward.

94 Yellow Shell
Camptogramma bilineata
Ws 21–27 mm   June–July
Common; Meadows, fields

This species occurs in large numbers in grassland and is active day and night. It is conspicuous and is easily distinguished by its yellow colour. The caterpillar is green with a darker line on its back. It feeds on chickweed, docks and grasses and, like many other geometrids, can assume the appearance of a twig by gripping a stalk with its abdominal prolegs and holding its body out stiffly at an angle (see also No 180). Hibernation is in the caterpillar stage.

95 Large Twin-spot Carpet ♂
Xanthorhoe quadrifasiata
Ws 22–30 mm   June–July   Locally common; meadows, glades, gardens

This very variable species occurs only in the southern half of England. The caterpillars feed on bedstraw, groundsel and other plants from August to April. They are typical loopers with grey markings on a yellowish-brown background. The pupa is found below ground.

96 Geometrid moth
Lythria purpuraria
Ws 19–23 mm   May–Aug   Dry grassland, heaths

This moth is not British although it has on at least one occasion been recorded in Scotland. Elsewhere in Europe it is common in dry, sunny habitats. It is on the wing in daytime visiting flowers and grasses. The markings of the wings are variable, completely red and completely brown individuals occurring. However, the kind illustrated here is commonest. The caterpillar is easily recognized. It is reddish on the back and yellow-green underneath. It lives on sorrel, especially sheep's-sorrel. It hibernates in the pupal stage.

97 Pyralid moth
Pyrausta purpuralis
Ws 13–18 mm May–Aug Common;
in meadows

This moth flies by day and night and
can often be seen visiting flowers
in sunshine. The caterpillars spin
together leaves of mint, thyme, or
plantain to form a shelter in which
they live. Two generations usually
occur during summer, the second
pupates during autumn and hibernates in a silk cocoon. Several
relatives are pests, e.g. the "mill-
moth" *Anagasta kuehniella* lives on
flour and grain in mills and the
"cacao-moth" *Ephestia elutella*
attacks cocoa, tobacco, and fodder
cake.

98 Grass-moths
Crambus species
Ws 17–27 mm May–Aug Common;
in grassland

These moths occur in large numbers
in grassland during summer. When
disturbed, they make short flights
before settling, often head downwards, and crawling down the stems.
The caterpillar makes a vertical,
silk-lined tunnel in the soil and
from this feeds on the roots and
bites through the stems of plants at,
or below, soil level. Sometimes large
areas of agricultural grasses are
killed and cereal crops may also be
affected. Usually the caterpillars
hibernate and pupation occurs in
spring. There are nine species in
Britain.

99 Plume-moth
Stenoptilia species
Ws 18–25 mm June–Sept
Common; meadows, woodsides

These do not look like ordinary
moths as their wings are divided into
narrow, fringed lobes and their legs
are long and spurred. They are most
active at dusk and during the night
but will fly by day if disturbed. At
rest they hold their wings and hind
legs outspread. The caterpillars feed
on many different plants. They hibernate in an early instar and pupate in
spring. The plate shows two individuals mating the female being
uppermost.

100 Swallow-tail (caterpillar)
Papilio machaon
L 40–50 mm   May–Aug   Very local; on umbels

The "swallow-tail butterfly" (No 1) lays eggs on umbels such as carrots, fennel, angelica and milk parsley. The young caterpillar is black with a white dorsal patch and resembles a bird dropping. Later it becomes yellowish-white with black and yellow cross-stripes and finally comes to resemble the plate. At this stage it develops a defence mechanism, an evil-smelling, red, fork-shaped structure which can be protruded from behind the head when it is alarmed. The pupa is attached to the host plant by a silken girdle.

101 Peacock (caterpillar)
Inachis io
L 40–50 mm   June–July   Common; on nettles

The "peacock butterfly" (No 11) lays eggs in large batches on the under-side of nettle leaves. The caterpillars remain close together throughout their development. They spin a web between the leaves and the stalk of the plant which is used as a kind of tent during moulting. When the nearby leaves have been consumed the caterpillars move on, leaving the web containing their cast skins, and make a new one. When fully grown, they move into the lower layers of the vegetation, attach themselves head downwards and pupate.

102 Elephant Hawk-moth (caterpillar)
Deilephila elpenor
L 60–80 mm   July–Aug   Common; on willow herb, marsh bedstraw, enchanter's nightshade

At first light-green, most individuals of this caterpillar become brownish-black and develop eye-like spots on the two segments behind the thorax. If it is disturbed it can withdraw the first three body segments into the fourth which bears the foremost pair of eyespots. When it does, these enlarge so that it appears menacing; presumably this scares potential predators. Pupation takes place in the organic litter on the ground. (The caterpillar of the "small elephant hawk-moth" is similar except it lacks the anal horn).

103 Fox-moth (caterpillar)
Macrothylacia rubi
L 60–80 mm   Aug–Oct   Common; grassland

This caterpillar hibernates full-grown. It wakens in early spring and, spinning a long brown cocoon, pupates on or near the ground. The moth emerges in May. Females fly at night but males are active by day searching for females in the vegetation. The caterpillars feed on grasses, blackberry, rose and heather and in late summer or autumn may be common on grassland. The hairs covering the body break easily if the caterpillar is handled and can cause a nettle-like rash on the human skin.

104 Bag-worm moth (larval case)
Psyche casta
L 12–14 mm   All year   Common; on vegetation, walls etc.

Bag-worms are so-called because the caterpillars spin cases round their bodies. The cases are covered with various scraps of vegetation and open at the rear to allow discharge of the faeces. The caterpillar always stays in the case which is enlarged when necessary. On reaching maturity, which may take two years, the caterpillar pupates within the case. The adult male emerges but the female usually remains inside. The cases are fixed to trees and walls.

105 Broom-moth (caterpillar)
Mamestra pisi
L 30–40 mm   Aug–Sept   Common; bushes, herbs, grass

This diurnally-active caterpillar is conspicuous by virtue of its yellow stripes on a green, brown or occasionally, black background. It feeds on broom, bracken, sweet gale, bramble, rose and sallow and, as its name suggests, on the cultivated pea. The adult is nocturnal and flies in June–July.

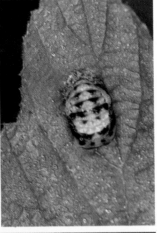

106 Seven Spot Ladybird
(larva and pupa)
Coccinella septempunctata
Larva 10–12 mm   June–Sept
Common; fields, gardens etc.

Ladybirds and their larvae are useful
insects. They feed on aphids and
scale-insects and help to keep their
numbers in check. The larva goes
through four developmental stages
the time taken being dependent on
temperature and the availability of
food. When full-grown the larva
attaches itself by means of a secretion to the substrate and pupates.
After about a week the adult
(No 107) emerges. Initially it is
completely yellow, but gradually
acquires its red colouration and
black spots.

107 Seven Spot Ladybird
Coccinella septempunctata
L 6–8 mm   All year   Common;
fields, gardens etc.

Of our 45 species of ladybirds this is
probably the commonest, at least in
rural districts. It hibernates, often
gregariously, under bark or in other
sheltered places, in the adult stage.
It reproduces very rapidly and sometimes occurs in vast numbers. Development from the egg to adult can,
under favourable conditions, take
less than one month, and several
broods can be produced in a summer.
The lady referred to in ''ladybird''
is the Virgin Mary who was often
depicted in red.

108 Twenty-two Spot Ladybird
Thea vigintiduopunctata
L 3–5 mm   June–Sept   Common;
on grass and herbs

This little ladybird is also brightly
coloured. Yellow, orange, and red
seem to serve as warning colours,
deterring or warning enemies of the
animals' disagreeable taste. If
directly handled ladybirds often
exude some of their body fluids
(reflex bleeding) so as to leave no
doubt in the predator's mind that
the animals are obnoxious. In the
past, this liquid was prized for its
analgesic qualities and was used to
relieve toothache.

109 Garden Chafer
Phylloperttha horticola
L 8–12 mm   June–July   Common;
grassland, bushes

The "garden chafer" is active by
day and can often be seen in mid-
summer swarming in large numbers
on grassland. The adult beetles eat
foliage of bushes, rose petals and
bracken but do no serious harm.
They lay their eggs in the soil and
the larvae feed on roots, growing
quickly to maturity in autumn. They
can cause serious damage to grassland, particularly on well-drained
sandy soils, and they also damage
cultivated crops.

110 Click-beetle
Corymbites cruciatus
L 11–14 mm   May–July   Meadows,
deciduous forest

There are over 8,000 species of
click-beetles. Their name derives
from the loud clicks the inverted
beetles make when righting themselves. On the under-side of the
first thoracic segment is a peg-like
process which fits into a socket in
the second. The peg is first with-
drawn under tension and then replaced rapidly with such force that the
beetle leaps into the air and lands
right way up. The species shown
here does not occur in Britain but
other species of *Corymbites* do.

111 Click-beetle
Agriotes aterrimus
L 12–15 mm   May–June   Meadows
in woodland areas

This non-British species is representative of a group of notorious agricultural pests. In particular, *A.
lineatus* and *A. obscurus* which are
slightly smaller and broader are
major pests. The adult beetles lay
their eggs in the soil and the larvae
(No 273), the well-known "wireworms" live for several years in the
ground eating roots and root crops,
especially potatoes. They pupate in
the late summer and after several
weeks the adults emerge to hibernate
in the soil. Another common name for
click-beetles is skipjacks.

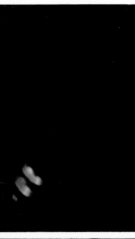
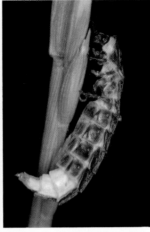

112 Glow-worm ♀
Lampyris noctiluca
L 11–18 mm　June–Aug　Common;
hedges, ditches

This is an adult female beetle though,
lacking elytra and wings, it looks
like a larva. The male is an ordinary-
looking beetle. On warm summer
nights the female sits with her abdomen raised, emitting a pale green
light (paler than in plate) to attract
the male. The light is produced by
the oxidation of a chemical, ''luciferin'', in the last three segments.
The light can be switched off if the
''glow-worm'' is disturbed. The
larva is predatory on snails. During
daytime these animals hide under
stones etc.

113 Oil-beetle ♀
Meloe violaceus
L 13–32 mm　April–June
Meadows, woods

This species is of rather local occurrence in Britain. When alarmed, it
exudes a stinking, yellowish, oily
liquid; an example of the reflex
bleeding seen in other beetles. The
egg-carrying females are large and
clumsy. Neither they nor the males
can fly. The eggs, several thousand
from each female, are laid in the
ground close to flowers. The larvae
climb into the flowers and attach
themselves to visiting bumblebees.
Further development takes place in
the nests of bumblebees where the
larvae feed on pollen and honey.

114 Vine-weevil
Otiorrhynchus sulcatus
L 8–11 mm　May–Sept　Common;
grassland, gardens etc.

The Latin name of this genus refers
to the ear-like extensions at the base
of the antennae. The majority of
species lack males and reproduction
is parthenogenetic i.e. without
fertilization. The larvae feed on roots
sometimes causing damage to strawberry plants and ornamentals. They
also occur in greenhouses, feeding
on the roots of cyclamen and ferns.
The adult is lethargic and lives
mainly on the ground. It can, however, chew holes in leaves of garden
flowers, such as rhododendron.

115 Leaf beetle
Chrysolina geminata
L 6–8 mm   On St. John's wort

This species does not occur in Britain although it is found on the Continent. Members of the genus are herbivores, each species being confined to one or a few closely related plants. The larvae feed on the same plants. A British species *Chrysolina varians*, also found on St John's wort, has been used successfully in the biological control of this plant in N. America. In some related forms the larvae develop inside the egg and hatch at the same time as, or before they are laid.

116 Tortoise-beetle
Cassida viridis
L 7–10 mm   June–Aug   Local; meadows, woodsides

This beetle thrives in moist habitats where it feeds on hemp-nettle and mint. If overturned, it can right itself by spreading out its elytra. The larvae are broad and flat and have spines all round the body. Faeces and cast larval skins collect on those of the abdomen and form a "case" which is carried by the larva. A related species, *Cassida nebulosa,* the "cloudy tortoise-beetle", is a serious pest of sugar beet on the Continent although it is rarely of consequence in Britain.

117 Cuckoo-spit insect (nymph)
Philaenus spumarius
Adult L 5–6 mm   May–Sept
Common; on herbs

The adults of these are small mottled insects which jump in the grass like miniature grasshoppers. Even if one fails to notice them the drops of "spit" (left) on plants are familiar. The spit is produced by the nymphs (right) which can easily be removed. They suck out plant sap which they mix in their alimentary canal with air and a surfactant. It is then passed out of the body to form the foam. The spit is an effective protection against predators.

118 Marsh Damsel-bug
Dolichonabis limbatus
L 7–10 mm May–Nov Common; damp meadows and fields

Damsel-bugs are predators of other insects. The bug seizes its prey in its front legs, pierces it with its mouth-parts and injects saliva into it. This paralyses the insect which is then sucked out. Individuals of this species usually have reduced wings, but fully winged specimens, all females, also occur. They lay eggs in vegetation and these hatch in the following spring. Close relatives in warmer climates suck the blood of birds and mammals. They also transmit disease.

119 Meadow Plant-bug ♂
Leptopterna dolabrata
L 8–9 mm May–Sept Common; grassland and corn fields

This bug feeds on grasses and is often abundant. It can damage cereals and in America is a major pest of meadow grass. The eggs are laid in summer at the base of grass stems and pass the winter in this stage. In spring, the larvae hatch and the first adults, usually males, are seen around the end of June. The females, which are greenish, usually have reduced wings and cannot fly, Like their relatives the shield-bugs, they have stink glands. Despite this they often fall prey to spiders.

120 Sloe-bug
Dolycoris baccarum
L 9–12 mm May–Sept Common; on shrubs and herbs

The shield-bugs, to which this belongs, are generally large, cumbersome, insects that secrete noxious fluids from their stink glands. Since these insects feed on berries in gardens and woods, this defensive propensity is well known to the layman. The adults of this species overwinter, and eggs are laid in batches on leaves in the spring. There are five larval instars before the adults appear in mid-August. The ''sloebug'', which is mainly a plant feeder, also does some good by eating aphids and beetles' eggs.

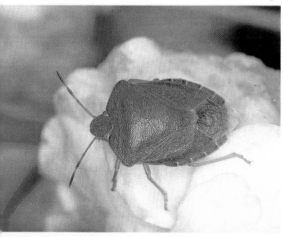

121 Green Shield-bug
Palomena prasina
L 11–14 mm   May–Nov   Common;
on herbs, bushes and deciduous trees

This bug can change colour. Bright
green in summer, it becomes dark
bronze-red before hibernating as an
adult. In spring it regains its green
colour. It feeds on many different
plants and on the continent is a
minor pest imparting an unpleasant
taste to raspberries. At first glance,
shield-bugs might be mistaken for
beetles, but are readily distinguished
by the presence of a rostrum and by
the nature of the fore-wings which
are not uniformly hardened like the
elytra of a beetle.

122 Ground-bug
Lygaeus equestris
L 7–12 mm   April–Sept   On dande-
lion

This handsomely marked bug receives its specific name in Latin from its
resemblance to a heraldic emblem.
In Western Europe it has a single
generation a year and hibernates as
an adult. This sun-loving species
occurs only accidentally in Britain.
Another species of somewhat similar
appearance, the ''firebug'', *Pyrrhoco-*
ris apterus, is, however, established
in Britain but at one locality only, a
rocky island off the coast of Devon.

123 Dusky Cockroach ♂
Ectobius lapponicus
L 7–10 mm   April–Oct   Local in S.
England; grass, scrub etc.

The cockroaches are a very old group
of insects, having existed for over
300 million years. Most cockroaches
in Northern Europe are introduced
species more or less confined to
buildings. The ''dusky cockroach''
is, however, native. In sunshine it is
a lively, fast-moving creature and
can be found basking or running over
fallen leaves. The female has reduced wings and cannot fly. Like other
cockroaches it lays eggs in a capsule
which is carried for a time attached
to the abdomen.

124 Wart-biter ♂
Decticus verrucivorus
L 31–37 mm   May–Sept   Rare;
heaths, grassy places

The "wart-biter" feeds on small
animals and plant materials. The
name comes from the old peasant
belief that warts would disappear if
it was made to bite them. The female
has a long ovipositor with which she
places eggs in the soil. There they
hibernate. The song of the "wart-
biter" is similar to that of the "great
green bush-cricket" (No 205) and is
produced by rubbing the adapted
wing bases together. Females are
attracted by the song: their "ears"
are on the fore-legs and resemble
dark cavities.

125 Bog Bush-Cricket ♂
Metrioptera brachyptera
L 11–21 mm   May–Nov   Common;
heaths, moors, bogs

This species is not easy to see as
it dives into dense undergrowth at
the approach of danger and is well
camouflaged with its green and
brown colouration. Some individuals,
however, lack the green colour on
the upper side. The song of this
species may persist all day when it
is sunny. It consists of short, shrill
chirps, repeated at intervals the
length of which depends on the temperature. Song diagram:

```
||||||||||||||||||||||||||||||||||||||||||||  time
0         5         10        15 sec
```

126 Dark Bush-cricket ♂
Pholidoptera griseoaptera
L 13–20 mm   April–Nov   Common;
roadsides, scrub, edges of woods

This species stays almost continuously in undergrowth close to the ground
where it is well concealed because
its colour matches that of its background. The song, mainly heard in the
evening, consists of short metallic
chirps and the singing of one male
often stimulates others nearby to
join in. It is produced by rubbing the
short wing stumps together. The female is completely wingless. She
lays her eggs in cracks in tree trunks
and the larvae hatch the following
spring. Song diagram:

```
|      |      |      |        time
0      5      10      15 sec
```

127 Short-horned Grasshopper
Psophus stridulus
L 20–34 mm   July–Sept   Dry, sunny
woodland slopes, heaths, etc

Grasshoppers are exclusively herbivorous and feed mainly on grass. This
species is not British but several related species are. It is well camouflaged when it sits in the grass but
when it takes flight its red hind-
wings become conspicuous. When
flying it makes a rattling noise like
a distant haymower. The dreaded
locusts of the tropics and subtropics
are species of grasshoppers which
have acquired the habit of forming
huge swarms which can cause devas-
tation to crops and vegetation.

128 Common Green Grasshopper ♀
Omocestus viridulus
L 14–22 mm   April–Oct   Common;
meadows, roadsides etc.

An extremely common grasshopper,
particularly where vegetation is lush.
Like other grasshoppers it is active
in the middle of the day. The female
is less active than the male and is
difficult to discover because of her
green colouration which matches her
surroundings. The male is smaller
and more variable in colour, greenish
and brownish forms being the most
common. The male hops and flies
readily if disturbed. The song consists of a long, loud ticking. Song
diagram:

129 Common Field Grasshopper ♂
Chorthippus brunneus
L 15–25 mm   May–Nov   Abundant;
roadsides, farmland

This species, like most other grasshoppers, is extremely variable in
colour. Brown and grey-mottled individuals are commonest. The males
sing actively in warm sunny weather.
The song consists of a sequence of
short isolated chirps produced by
the inside of the spined femur of the
hind-leg being rubbed against the
wing. The ''ear'' is situated on the
second abdominal segment. Grasshoppers lay their eggs in the ground
and cover them with a secretion
which hardens to form a protective
capsule. The larvae hatch in spring.
Song diagram:

130 Ichneumon-fly
Enicospilus merdarius
L 15–20 mm June–Sept Common; meadows

This parasitic wasp may be seen flying over grassland searching for hosts such as caterpillars of hawk-moths and noctuids. Eggs are laid in the hosts and the larvae develop within them. At first they feed on fluids only and cause little harm to their hosts but later they start to eat the tissues and the hosts become moribund and die. The parasite larvae then leave their hosts and pupate. Parasitic wasps are important in regulating the populations of many species of insects.

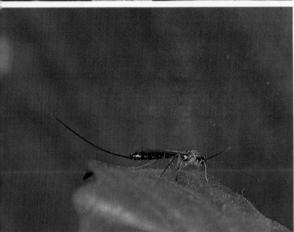

131 Ichneumon-fly ♀
Perithous divinator
L 10–17 mm June–July Common; in meadows and woods

The females of some parasitic wasps have long ovipositors with fine saw-like points. By means of these they penetrate plant tissues, sometimes even hard wood, and lay eggs in insect larvae living inside. The antennae of these ichneumonids are enormously sensitive and can detect the presence of concealed hosts. The species illustrated is a parasite of predatory wasps, whose larvae live in the stalks of raspberry and hops. Parasitic wasps are cultivated for use in controlling crop pests, as an alternative or complement to pesticides.

132 Saw-fly
Tenthredo mesomelas
L 10–13 mm May–July Common on herbs and bushes

Saw-flies are herbivorous wasps. The females possess saw-like ovipositors and make grooves in stems and stalks. One egg is laid in each groove, a female laying about 50 in all. The larva of the species shown eats the leaves of various plants, including buttercups and speedwells. Adult saw-flies usually live for a few days only. Some prey on flies etc. but more commonly they eat pollen from flowers, especially umbels.

133 Crane-fly ♀
Tipula maxima
L 30–40 mm   April–Aug
Frequent; in woodland

All crane-flies have legs about twice as long as the body and are known as daddy-long-legs. They are clumsy walkers and one usually sees them either sitting on vegetation or flying. If caught, they readily shed legs with no apparent ill-effects. Like all true flies, craneflies have only one pair of functional wings, the second pair being modified to balancing organs. The female lays eggs in moist ground. The larvae live in the mud of shallow waters and eat rotting vegetation.

134 Common Crane-fly ♀
Tipula paludosa
L 17–25 mm   July–Sept   Common; agricultural land

This crane-fly is common on all grassland. The female performs a peculiar hopping and dancing flight during egg-laying. Each time she makes contact with the ground an egg is laid. The larvae live in the soil on roots and rotting vegetable matter and seriously impair the productivity of grassland. They also damage arable crops such as cereals. Sometimes, after dark, the larvae come to the surface and bite through the seedling plants. They pass winter in the soil, pupate in spring and the adults appear during summer.

135 Crane-fly
Nephrotoma flavescens
L 14–20 mm June–Sept   Common; woodland, scrub, gardens etc.

A male and female *in copula* can remain firmly attached for a considerable time and can often be found sitting in the vegetation. In the plate the male is to the left, the female to the right. *Nephrotoma* species rarely, if ever, cause damage to grassland and crops in Britain, but in parts of the continent they are said to injure the roots of young coniferous trees.

136 St Mark's-fly ♀
Bibio marci
L 9–14 mm   March–June   Common; fields, gardens

This species is called ''St Mark's-fly'' because it appears in numbers about 25 April, St Mark's day. It flies in a slow, oscillatory manner with its legs hanging down. Mating takes place in flight, the female flying up and the male catching hold of her. Eggs are laid in the soil and the larvae feed on various roots and decaying vegetation. Winter is spent as a larva and pupation takes place in early spring. The species is probably of little or no economic importance in Britain.

137 Sepsid-fly
Sepsis cynipsea
L 4–6 mm   April–Aug   Common; grassland, scrub

These ant-like flies often occur in large swarms moving around over the vegetation. They are characterized by their habit of wing-waving and by the dark spots on the tips of their wings. Their larvae feed on rotting organic matter and the flies are often encountered near compost and dung heaps.

138 Helomyzid-fly
Helomyza fuscicornis
L 4–6 mm   June–Nov   Common; in shady places, woods etc.

These flies frequent dark damp places, particularly where there are fungi growing or rotting plant remains. Their larvae develop within toadstools. There are many insects, mainly flies and midges, whose larvae share this habitat. Usually, they eat their way up the stalk and then attack the cap. Many species, however, lay eggs directly in the vanes or pores on the under-side of the cap. To identify the species in a fungus it is usually necessary to rear the adults.

139 Robber-fly
Dysmachus picipes
L 10–16 mm   June–Aug   Meadows, glades

Robber-flies capture and eat other insects. They often sit still on the ground or on a stone and wait. When insects approach, they fly up and catch them in mid-air grasping them with their legs. They lay eggs on grass. After hatching, the larvae crawl into the ground where they feed on plant material. Robber-flies are characterized by their broad heads and large protruding eyes. They have sharp vision, which is necessary for their way of life. The species shown does not occur in Britain.

140 Robber-fly
Dioctria rufipes
L 11–14 mm   May–July   Common; meadows, glades

This robber-fly lives primarily on small parasitic wasps. It does not lie in wait for prey but crawls about in the vegetation. When a suitable prey appears, it takes off and catches it in flight. The prey is killed by the injection of poison. In flight this species holds its fore-legs out in front and the hind-legs hang down from the body. Mating couples may often be seen in vegetation. Like all robber-flies, it holds its wings folded over the abdomen when resting.

141 Robber-fly
Leptogaster cylindrica
L 7–15 mm   May–Aug   Common; meadows, woodland

This small, slender insect flies slowly through grass and round bushes in search of suitable prey. It specializes on leaf hoppers, aphids and small two-winged flies. It often hangs on to a blade of grass with one leg while clasping its prey with the others. During oviposition the female sits on the grass and lets her eggs fall to the ground. The larvae feed mainly on rotting plant remains, but are thought to attack larvae of other insects as well.

142 Garden-spider (web)
Araneus diadematus
July–Oct Common; grassland, woods, walls

Spiders have silk glands and spinnerets. Silk is secreted as a liquid which solidifies by reorientation of its molecules as it is drawn out. The garden spider is an orb-web builder. It makes its wheel-shaped web by first dropping a thread from a high point so that it catches onto a twig. It then secures this firmly at both ends. Next, it weaves the outer margins, the radial threads and the centre. The spiral of sticky thread, which actually catches the prey, is woven onto this foundation.

143 Garden-spider ♀
Araneus diadematus
L 7–18 mm July–Oct Common; grassland, woods, walls

The last thread to be spun starts from the middle of the web and leads to the spider's hiding place. When an insect gets caught, vibrations are conducted to the spider. The male is considerably smaller than the female, a common condition in spiders. Mating takes place in late summer. The male diffidently approaches the female and if she is receptive, mating takes place. Afterwards he has to retreat quickly or she will eat him! The female lays her eggs in a yellow silken cocoon, 400–800 in all.

144 Orb-web building spider ♀
Tetragnatha extensa
L 6–11 mm June–Oct Common; in low vegetation, especially near water

These spiders weave webs low down in vegetation e.g. between reeds or in low bushes. The spider sits in the middle waiting for prey. (The plate shows a female hanging under her web). When these spiders rest they are easily recognized by the way the long fore-legs are held out in front of the body. The powerful spined jaws (chelicerae) of the male are characteristic. He uses them to hold those of the female during mating so that she cannot attack him.

145 Comb-footed spider
Theridion ovatum
L 3–6 mm   June–Oct   Common; on undergrowth, nettles etc.

The spherical abdomen is a characteristic of comb-footed spiders. They feed on various small insects which are caught in their irregular, crisscross webs. The females construct nests and place their egg-cocoons in rolled-up leaves. In some species the females regurgitate liquid food for their young but it is more usual to provide them with prey. The young stay in the nests until they are quite large. The species illustrated occurs in two colour forms; brownish-yellow and crimson-striped.

146 Crab-spider ♀
Micrommata virescens
L 9–14 mm   June–Sept   Local, mainly in South; woods

This spider is very fast and catches prey by jumping on it. It has tufts of hairs on the ends of its legs which act as adhesive surfaces when it jumps about in vegetation. The male is yellow with three red longitudinal bands along the abdomen. During mating he grasps the female from behind and jumps on her back. As in other spiders, he transfers sperm prior to mating to the swollen and complicated end segment of his pedipalps which he uses as a copulatory organ.

147 Wolf-spider ♀
Pisaura mirabilis
L 8–14 mm   May–Sept   Common; dry grassland, heaths

These spiders do not catch their prey in webs but run after and jump on them. Prior to mating the male offers an insect swathed in silk to the female. She eats this and mates at the same time. A male attempting copulation without such a present would usually get eaten. The female spins a cocoon for her eggs. She carries this with her in her jaws (plate) and weaves a kind of nursery where she places it before hatching. She guards the nursery when the young have hatched.

148 Roman Snail
Helix pomatia
H 42–45 mm   April–Oct   Local;
woods, scrub etc. on calcareous soil

The largest British snail. It occurs
throughout most of Europe and seems
to be native in the places in southern
England where it occurs. Some bio-
logists think, however, that it was
introduced to Britain by the Romans
as a culinary delicacy. Like many
other terrestrial snails it is herma-
phroditic, i.e. the same individual is
both male and female. Copulation is
reciprocal (plate), and during its
course the pair shoot each other with
tiny dart-like projectiles which penetrate deep into their bodies. These
are supposed to be sexually stimula-
ting.

149 Smaller Banded-snail
Helix hortensis
H 12–15 mm   May–Sept   Common;
gardens, scrub etc.

This snail varies greatly in appearance. It can be either yellow or pink,
and with or without longitudinal
bands. Moreover, any one or more of
the five bands may be lacking, or be
fused with others. A close relative is
H. nemoralis, the ''larger banded-
snail'', which is similar and equally
polymorphic but has a dark-lipped
shell. Both species hibernate in the
ground and become quiescent when
the weather is dry. They feed on
fresh and rotting plant materials.

150 Slug
Agriolimax agrestis
L 30–50 mm   June–Nov   Local;
fields

Slugs, like terrestrial snails, are air
breathers. The mantle cavity serves
as a lung and its opening as a respiratory pore. This species is a comparative rarity in Britain which for
many years was confused with the
common and ubiquitous ''field-slug''
A. reticulatus, which is a notorious
pest of many crops. Slugs have file-
like rasping tongues and are most
active after dark, moving best on
damp surfaces. The ''field-slug''
lays up to 300 eggs in the soil in
batches of up to 30 at a time.

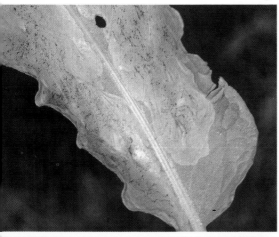

**151 Blister mine on dock, mining
fly larvae**
Pegomya species
Larva L 5–8 mm   May–Oct
Common; on dock leaves

These flies lay eggs on leaves of
various species of dock. After
hatching, the larvae gnaw through
the upper surface of the leaf and
start to eat the tissues. They do not
eat the cuticle and this remains as a
thin, transparent membrane through
which the larvae can be seen (plate).
The mine starts off narrow and
linear but soon becomes broad and
blister-like. It usually contains
several larvae. Finally, the entire
leaf turns brown and dies. The adult
fly is about 6 mm long.

**152 Linear mine on Golden-rod,
mining fly** (larvae)
Ophiomyia maura
Larva L 4–5 mm   June–Aug   Local;
on leaves of Golden-rod

These mines are common on plants
such as thistle and plantain. They are
formed when the larvae of certain
beetles, sawflies, moths or flies
feed on the leaf tissues between the
upper and lower epidermis. It is
possible to identify the insect responsible by knowing the host plant
species, the form of the mine and
and the position of the excrement
within it. The species shown here is
only one of several hundred mining
flies found in Britain. It occurs in
parts of southern England and the
Midlands.

**153 Leaf gall on Common
speedwell, gall midge** (larvae)
Jaapiella veronicae
Larva 3–4 mm   On Common speedwell

This gall midge lays eggs in a growing bud of common speedwell. The
bud swells up into a spherical structure consisting of thickened, deformed leaf-pairs which become covered
with a fine, white pubescence. These
changes are due to chemical substances produced by the midge
larvae living inside the gall. Later
adults emerge from the galls as
minute winged midges.

154 Green Hairstreak
Callophrys rubi
Ws 23–25 mm   May–June
Common; woodsides, clearings

This lively butterfly is often seen at the edges of woods and in clearings. At rest, the wings are always closed and it is then well camouflaged as it sits on the vegetation – the upper sides of the wings are brown. The host plants of the caterpillars are furze, bird's-foot trefoil and rock-rose. Pupation occurs in August and the pupa hibernates. There are five species of hairstreaks in Britain. They can be distinguished from other butterflies by the small projections on their hind-wings.

155 Privet Hawk-moth (caterpillar)
Sphinx ligustri
L 80–100 mm   July–Aug   Common in South; privet, lilac, ash

On hatching, the young caterpillar of this species eats the eggshell and then starts to feed on the leaf it is on. After 4–5 moults it is full-grown and as thick as a finger. Like most hawk-moth larvae it has a large anal horn. If in danger it assumes a characteristic raised posture like that of the sphinx (hence the Latin name). Before pupating it digs a hole about 10–15 cm down in the ground wherein the pupa hibernates. Adult No 156.

156 Privet Hawk-moth
Sphinx ligustri
Ws 85–125 mm   June–July
Common in South; gardens, parks etc.

This large hawk-moth can be seen on summer evenings visiting the flowers of its host plants. It can hover in the air in front of a flower with its proboscis inserted into the corolla. Later in the evening the moths perform mating flights in which several males swarm around a female until one of them succeeds in mating with her. Like most night-flying moths, this species is readily taken at light. An even larger species, the "death's head hawk-moth", occurs only sporadically in the British Isles.

157 Eyed Hawk-moth
Smerinthus ocellata
Ws 65–90 mm   May–July   Locally common; gardens, parks etc.

This species has large eye spots on the hind-wings. At rest these are covered by the fore-wings and the insect resembles a dead leaf. If disturbed, however, it reveals its "eyes". It flies in the evening and at night, but unlike most hawk-moths does not visit flowers. The male has an ephemeral existence and dies within 3–5 days. Its sole function is to find a female and fertilize her. Eggs are laid on sallow, aspen, poplar and occasionally fruit trees. Caterpillars inhabit these in July–August.

158 Poplar Hawk-moth
Laothoe populi
Ws 57–90 mm   May–Aug   Common; gardens, park etc.

Like the previous species, this does not visit flowers. Its nocturnal flight is exclusively for mating. Male and female remain *in copula* for up to 24 hr. (In the plate the female is uppermost). At rest, it holds its hind-wings in front of the fore-wings – a very characteristic posture. The caterpillar is green with a yellow oblique stripe and an anal horn. It feeds on poplar, aspen, willow and sallow from July to September, and grows to 70–80 mm. The pupa hibernates just below the soil surface.

159 Vapourer (caterpillar)
Orgyia antiqua
L 25–30 mm   May–Aug   Common; bushes etc.

Tussock-moth caterpillars bear tufts of hair on the back and at either end of the body. That of the "vapourer" is common on many types of tree and shrub including oak, lime and hawthorn. The adult male, which is brownish, can be seen by day on the wing in search of females. The female is a wingless, greyish-brown, clumsy creature which remains clinging to the pupal cocoon. She lays eggs in or around the cocoon then falls to the ground and dies.

160 Puss-moth (caterpillar)
Dicranura vinula
L 60–70 mm   July–Sept   Common;
on aspen, poplar, sallow

This caterpillar has a very characteristic pair of tail-like anal styles.
The first instar larva is black-brown and resembles a tiny slug. With each moult the colour becomes increasingly greener until it attains the appearance it has in the plate. If molested, it shoots out a red thread from each anal style. At the same time the thorax swells and two false eyes are displayed. It also ejects fluid containing formic acid which both burns and causes irritation.
A well-defended creature! Adult No 161.

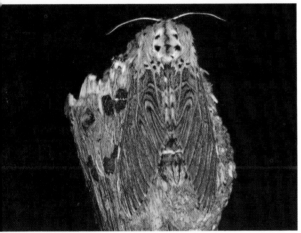

161 Puss-moth
Dicranura vinula
Ws 45–75 mm   May–July
Common; in woodland

Prior to pupation the caterpillar constructs a hard cocoon from pieces of bark in a specially bored-out chamber in a bough of a tree. In this situation the pupa is well protected and can remain dormant for several years. Before emerging, the moth secretes a liquid which softens up the wall of the cocoon so that it can get out. The plate shows a newly emerged moth, sitting on its cocoon. It flies only during the night remaining hidden in daytime. Caterpillar No 160.

162 Gypsy-moth (caterpillar)
Lymantria dispar
L 40–70 mm   May–June   On fruit trees and other deciduous trees

In parts of E. and C. Europe and of N. America this species is a pest. In Britain it used to be common in E. Anglia but died out over 100 years ago. Isolated individuals, possibly migrants, still occur along the S. coast. The caterpillars are voracious feeders and can ravage large areas of woodland. They have been controlled in N. America by introducing the predacious groundbeetle, *Calosoma sycophanta,* and by using synthetically prepared female sex scent to attract males to their doom on a sticky trap.

163 Buff-tip (caterpillars)
Phalera bucephala
L 50–60 mm   Aug–Sept   Common;
on deciduous trees

A gregarious species in the larval
stages. The individuals of a brood
remain together until the last
moult. They eat simultaneously on
the same branch and all rest at the
same time. When they have stripped
one twig they go to another and
resume their activities. If molested
they raise both their back and front
ends. One should avoid touching
these caterpillars and their relatives
because they possess long, brittle
hairs which can penetrate skin and
cause severe nettle rash or even
eczema. Adult No 164.

164 Buff-tip
Phalera bucephala
Ws 45–65   June–July   Common;
deciduous woods, parks, gardens

This moth flies only at night; during
the day it rests on the ground or a
tree with its wings folded over the
body where it looks like a twig
(plate). When picked up it feigns
death and can be rolled in the hand
without eliciting a response – an
effective defence against predators
that take only live prey. The female
lays eggs in batches on foliage of
deciduous trees. They are hemispheri-
cal and light-coloured with a dark
spot which is the porous point
through which the sperms penetrate.

165 Garden Tiger
Arctia caja
Ws 43–70 mm   July–Aug   Common;
woods, meadows, gardens

There are several different tiger
moths, all strikingly coloured. This
species is a nocturnal flyer. At rest
the bright red hind-wings are
usually concealed, but if it is disturbed, they are displayed warning
intending predators that it is distasteful. It secretes a noxious fluid
from its thorax. Eggs are laid on the
under-sides of leaves of many
different plants and in S. Europe
caterpillars cause damage to vineyards. The caterpillar hibernates in
an early instar and pupates the
following summer in a cocoon on the
ground.

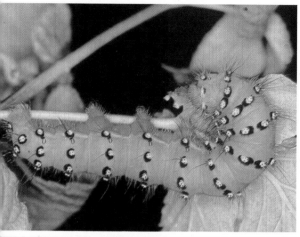

166 Emperor-moth (caterpillar)
Saturnia pavonia
L 50–70 mm   June–Sept   Locally
common; bramble, heaths etc.

The newly hatched caterpillar is
almost completely black. As it grows
it gradually turns green with yellow
or reddish, hairy warts. Full-grown
it spins a brown pear-shaped cocoon
that is covered with a loose web.
At the pointed end there is a hole
surrounded on the outside by a whorl
of stiff, outwards-pointing bristles to
hinder the entry of intruders. These
do not, of course, hinder the exit of
the moth. Hibernation takes place
in the pupal stage and the adult
emerges in the following spring.

167 Emperor-moth ♂
Saturnia pavonia
Ws 42–70 mm   April–May   Locally
common; heaths, moors

The male is most often seen as it is
active by day. The larger and paler
female rests in foliage during the
day but becomes active at dusk. The
male seeks out the female by means
of his antennae (see plate). Like
many moths the female produces sex
scents (pheromones) which attract
males from long distances. It has
been found that only a few molecules are necessary to stimulate flight
activity in the males. Attempts are
being made to use pheromones to
control lepidopterous pests e.g. No
162.

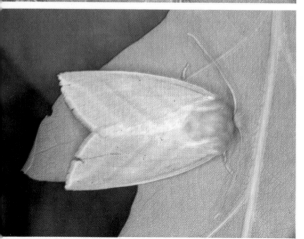

168 Green Silver Lines
Pseudoips fagana
Ws 30–38 mm   May–June
Common; deciduous woodland

This moth rests for most of the day
in trees or undergrowth and takes
wing when it gets dark. The female
is larger than the male. Her hind
wings are white, whereas those of
the male are yellowish. The caterpillar lives on the leaves of various
deciduous trees. It is green with
yellow spots and stripes, and about
30 mm long when full-grown. Prior
to pupation it builds a boat-shaped
cocoon on the ground or on a tree
trunk. Hibernation takes place in the
cocoon.

169 Common Footman
Eilema lurideola
Ws 25–35 mm   June–July
Common; deciduous and coniferous woods

This insignificant little moth is usually on the wing at night, but is often flushed from vegetation during the day. The female lays eggs in clusters on lichen-covered tree trunks or branches. The caterpillar attains a length of 25–30 mm and is black or greyish-blue with orange lateral stripes and black hairs on its back. It hibernates and later pupates in a cocoon among the lichens on a tree trunk.

170 Pebble Hook-tip
Drepana falcataria
Ws 27–35 mm   May–Aug   Locally common; in woodland

There are five species of hook-tip in Britain. They are characterised by the sickle-shaped, pointed forewing. This species is widespread and is locally common, especially on birches. The caterpillar is 25–30 mm long, brownish-red on top and yellowish-green at the sides. There may be two generations, caterpillars occurring in June–July and in Sept–Oct. The main food plants are birch and alder and the cocoon is made within a shelter formed from a folded leaf secured with silk. The pupa overwinters.

171 Goat-moth
Cossus cossus
Ws 65–88 mm   June–July
Common; in woodland

This large, clumsy moth flies only by night. In the day it rests on tree trunks, etc. the colours of which often match it closely. It gets its English name from the caterpillar which smells like a Billy-goat. It feeds on the wood of various trees – elm, ash, willow and birch – and can take several years to complete its development. The adult has rudimentary mouthparts and is unable to feed. The female lays its eggs in cracks of bark. Caterpillar No 271.

172 Red Underwing
Catocala nupta
Ws 67–78 mm June–July Local
in S. and E. England; woodland

One of our largest and most beautiful
of moths. In daytime as it sits on tree
trunks etc. it is well concealed due
to the markings of the forewings. The
hind-wings are brilliantly coloured
and are displayed if it is disturbed.
Eggs are laid singly on deciduous
trees e.g. poplar. They hatch the
following spring and the caterpillars
feed on foliage during the night.
They are brownish and have a row of
outgrowths on either side with which
they maintain close contact with
the substrate.

173 Large Yellow Underwing
Noctua pronuba
Ws 46–60 mm June–Oct Com-
mon; woods, gardens, fields

This largish noctuid visits flowers
and "bleeding" trees at nightfall
to obtain liquid food. It is also seen
if flushed from its hiding place during
the day. The female lays eggs in
clusters on blades of grass or other
herbs. The caterpillars, which have a
broad range of host plants, occasionally damage vegetables such as
lettuce and cabbage. Hibernation
occurs in the pupal stage. The colour
and markings of the fore-wings vary
greatly between individuals. There
are several rather similar species of
yellow underwings.

174 Knot Grass (caterpillar)
Acronycta rumicis
L 32–37 mm July–Sept Common;
roadsides, hedges etc.

Unlike other noctuids, the caterpillars of the subfamily to which this
species belongs are very hairy.
They feed on plants such as plantain,
dock, sorrel, hawthorn, bramble and
sallow. The moth is on the wing in
June and July, has a wingspan of
30–37 mm and a grey ground colour.
Its markings, which are black or
brownish, vary a great deal in con-
spicuousness.

175 Large Emerald
Geometra papilionaria
Ws 39–53 mm   June–July
Common; deciduous woodland

This large, green moth looks like a butterfly but is, in fact, a night-flying geometer. In the day it rests among foliage protected by its concealing colouration and is found where its foodplants, birch, hazel and beech occur. The female lays eggs singly on the under-sides of leaves and the half-grown caterpillar hibernates sitting on a branch. It is attached by its abdominal legs and the body is held out stiffly giving the appearance of a thorn. It pupates in spring in a spun-together leaf.

176 Brimstone-moth
Opisthograptis luteolata
Ws 26–42 mm   April–Aug   Common; woods, gardens etc.

A lively colourful moth that flies in twilight among trees and bushes often in gardens. It occasionally also flies on sunny mornings. The caterpillar feeds chiefly on hawthorn but also eats blackthorn, plum, rowan etc. It reaches a length of 25 mm and is brown or greenish with fringe-like projections on the sides of the last three segments. The pupa hibernates in a white cocoon among fallen leaves or in moss. A second brood appears in August–September.

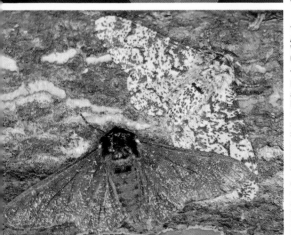

177 Peppered-moth
Biston betularia
Ws 32–55 mm   May–July
Common; deciduous woodland

The plate shows two extremes, a light and a dark form. In industrial areas this species has gradually become darker. This is because the whiter forms have become increasingly conspicuous to predators on sooty bark. Elsewhere the lighter animal matches the background of lichen-covered tree trunks. This is known as ''industrial melanism''. The female lays several hundred eggs. The caterpillars feed on various deciduous trees and bushes and vary in appearance to conform with the shoots of the food plant. Pupation occurs in a hole in the ground.

178 Common White Wave ♀
Cabera pusaria
Ws 25–34 mm May–Aug Common;
deciduous woods, swamps

This white moth usually bears three
grey lines across the fore-wings and
two across the hind-wings although
there are sometimes fewer. The
ground colour can also deviate from
white to greyish or pinkish. The caterpillar feeds mainly on birch or alder. It varies in colour being either
purplish-brown or green with a
brownish-red mark on the back. It
can attain a length of 25 mm. There
are two broods, one in July, the other
in September. The pupa hibernates in
a cocoon on the ground.

179 Clouded Border
Lomaspilis marginata
Ws 19–28 mm May–Aug Common;
moist deciduous woodland

This common rather attractive little
moth is on the wing both day and
night and can be seen throughout the
summer. It often sits with its wings
outspread on a leaf but despite
this can be difficult to detect because the dark markings obscure its
outline. It is, however, alert to
danger and is not easily approached.
The caterpillars become 18–20 mm
long and are dark green. They feed
on sallow, willow, aspen, etc. The
pupa hibernates within a silken
cocoon near the host plant.

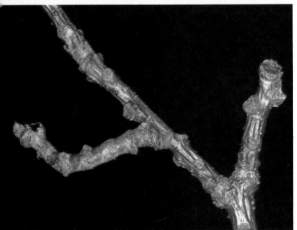

180 Thorn-moth (caterpillar)
Selenia species
L 35–40 mm May–Sept Common;
woodland, gardens

Where is the caterpillar? It bears a
close resemblance to the brown,
knotty twigs and if disturbed
becomes immobile and stick-like – a
highly effective protection from its
enemies. It is frequently attached to
the stem by means of a silken thread
in case it should fall. These caterpillars, (several different species)
can be found on various deciduous
trees – birch, sallow, hawthorn – and
on dwarf shrubs such as whortle-
berry. The pupa hibernates in a silken cocoon on the ground or in the
soil.

181 Mottled Umber
Erannis defoliaria
Ws ♂ 30–44 mm   Oct–March
Common; deciduous woods

This species, like the following one (No 182) shows a pronounced sexual dimorphism. The male (left) is a fully-winged typical geometer, somewhat variable in colour and markings, while the female (right) is wingless. The female crawls up trunks of deciduous trees in the autumn, mates and lays her eggs around buds. The caterpillars hatch in spring and commence feeding on the swollen buds and new foliage. In some years they occur in such numbers that whole trees are defoliated. They are reddish-brown and grow to 30–35 mm.

182 Winter-moth
Operophtera brumata
Ws ♂ 25–30 mm   Oct–Jan
Common; deciduous woods, parks, gardens

The female has rudimentary wings and cannot fly. In autumn, she crawls up the tree, mates with one of the males swarming in the canopy and lays eggs near buds. The green caterpillars hatch in spring and can cause extensive damage to various deciduous trees, including fruit trees. The caterpillars mature by mid-summer and lower themselves to the ground where they pupate. To control them, paint the tree trunks with a special adhesive banding compound to catch the ascending females.

183 Green Oak Tortrix-moth
Tortrix viridana
Ws 18–22 mm   June–July
Common; oak woods

In this group of moths the larvae roll up leaves to form shelters. The female lays eggs in pairs on branches of oak trees and covers them with scales from her own wings mixed with a secretion. The eggs hatch in May–June. The small, green caterpillars can defoliate large stands of oak. If one stands in an infested wood one can actually hear the sound of myriads of caterpillars chewing away at the leaves and that of the excrement (frass) falling on the forest floor.

184 Many-plume-moth
Alucita hexadactyla
Ws 13–16 mm   Aug–April
Common; deciduous woods

This moth rests by day in vegetation often with its wings outspread like a fan. It flies at dusk, is attracted to light and comes to lighted windows on summer evenings. Each wing is divided almost to the base into 6–7 feather-like lobes. The moth has a long proboscis and is often found on wild honeysuckle. The caterpillar also lives on honeysuckle, feeding on leaves and flowers. It pupates in a spun cocoon on, or just below, the ground. The adult hibernates in sheltered places.

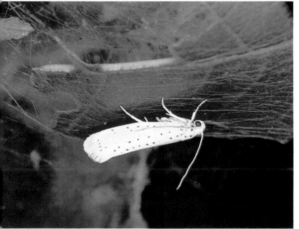

185 Small Ermine-moths
Yponomeuta species
Ws 15–25 mm   July–Aug   Common; deciduous woods, scrub, gardens

These caterpillars feed on leaves of bird-cherry, hawthorn, apple, spindle, blackthorn etc. They cover the leaves with a net of silk and feed and pupate within its shelter. Keeping close together, they move from branch to branch, covering themselves as they go. When in large numbers they can defoliate and even kill bushes. The eggs are laid and hatch in autumn but the small caterpillars remain within the egg cases until the following spring. The plate shows a newly emerged moth sitting on a web-covered leaf.

186 Sun-moth ♂
Nemophora degeerella
Ws 15–21 mm   June–July
Common; moist woodland

On sunny summer mornings males of these moths dance up and down under trees floating gently down ''parachuting'' with their enormously long antennae. The females have shorter antennae and while the males are swarming sit in vegetation nearby waiting for a mate. The caterpillars feed on leaves of birch and oak. In the autumn they drop to the ground with the leaves and there construct a shelter of leaf fragments. They pupate in the spring.
Adela reamurella, with green, metallic wings, is just as common in deciduous woodland.

187 Scorpion-fly ♂
Panorpa communis
L 18–22 mm   May–Sept   Common;
woods, gardens

The name of this insect refers to
the copulatory claw of the male that
is curled up over the abdomen like
the tail of a scorpion. This claw is
used to grasp the female during
mating. The male secretes small
drops of saliva which he feeds to the
female before and during mating.
The insect eats dead insects as well
as pollen and nectar. The eggs are
laid in the soil about 100 at a time.
The larvae hibernate in holes in the
ground and pupate in the spring.

188 Empid-fly
Empis tessellata
L 7–11 mm   May–July   Common;
scrub and woodland

This fly is an omnivore, feeding on
insects as well as on nectar. Sometimes one can be encountered flying
heavily laden with its prey, a gnat or
fly. Occasionally large numbers
occur on flower-heads of umbels.
Prior to mating, swarming occurs,
often in moist shady places. The
males of some *Empis* species give
the female a ''wedding present''
consisting of an insect wrapped up in
a silken thread. Mating takes place
while the female consumes it.
The larvae are predacious living in
the soil or in rotting wood.

189 Snipe-fly
Rhagio scolopacea
L 8–14 mm   May–July   Common;
woodland, meadows

Snipe-flies have a short, anteriorly
directed proboscis and have been
accused of preying on other insects.
Their feeding habits are, however
largely unknown though some species
are known to take mammalian blood.
R. scolopacea often sits head down
wards on a tree and flies round,
but never attacks humans. Its flight
somewhat resembles that of a snipe.
The larvae live in the soil and feed on
worms, insect larvae etc.

190 Stag-beetle ♂
Lucanus cervus
L ♂ 30–80 mm   June–Aug   Common
locally in S. England

Europe's largest beetle. It lives in old
woods and as these are successively
cleared, is becoming scarcer. In
Britain it is restricted to southern
England, on the Continent it is local
and sporadic, extending northwards
into southern Sweden. It feeds on
exuding sap and is confined to senile
or rotting trees. On warm summer
evenings males fly noisily in search
of females sitting in crowns of trees.
In flight the enormous jaws are held
upwards and the abdomen hangs
almost straight down. Males engage
in violent battles over females.

191 Stag-beetle ♀
Lucanus cervus
L ♀ 25–50 mm   June–Aug   Common
locally in S. England

The females are usually darker and
smaller than the males and do not
vary so much in size. After mating
they seek suitable places to lay
eggs, usually rotten oak trunks (occasionally beech). The larva lives in
the rotten wood for up to five years
before becoming full-grown (10 cm
long). Before pupating it constructs a
cocoon of rotten wood. The male larvae must make much larger cocoons
since they remain inside them until
the large jaws have attained their
final form and hardness.

192 Chafer-beetle
Melolontha hippocastani
L 20–30 mm   May–June   Local in
N. England and Scotland; woodland

These chafers fly on warm spring
evenings in the crowns of deciduous
trees where they eat the foliage. The
larvae feed on roots but are of no
economic importance. A near relative,
the "cockchafer", *M. melolontha*
occurs in the more southerly parts of
the country and is a pest as its larvae can cause serious damage to
grassland. The larvae take three
years to mature and if the adults are
unusually abundant one year then
damage due to the larvae is likely to
be serious two years later.

193 Summer-chafer
Amphimallon solstitialis
L 14–18 mm   June–July   Locally common; woodland

This species swarms among the tops of trees on warm summer evenings often in very large numbers. It is commonly confused with the ''cock-chafer'' or with *Melolontha hippocastani* (No 192) but is considerably more hairy. The beetle feeds on leaves and needles of trees and defoliates entire branches. Eggs are laid in the ground. The larvae (No 272) look like those of the ''cock-chafer'' but live in the soil for only two years where they feed on roots. The species has a rather sporadic distribution in the southern half of Britain.

194 Brown-chafer
Serica brunnea
L 8–10 mm   July–Aug   Local; sandy places

A nocturnal beetle that sometimes flies into houses during the summer, presumably attracted by light. It leads a retiring existence during the daytime keeping out of sight under stones, etc. Although widely distributed in Britain it is confined to sandy places, and consequently has a local distribution. The larvae feed on roots.

195 Cardinal-beetle
Pyrochroa coccinea
L 14–15 mm   June–July   Very local; deciduous woods

This beautiful beetle is confined in Britain to southern England and some parts of Wales, where it can be seen climbing on branches of trees or in flowers. The adult presumably lives for only a short period but the larvae, which live in rotten wood, take up to 3 years before pupating. This is thought to be due to the low nutritional value of the food. Another species of ''cardinal'' *P. serraticornis* has, unlike this species, a scarlet head.

196 Large Poplar Longhorn
Saperda carcharias
L 18–28 mm   June–July   Rare; on aspen, poplar, willow

This large beetle is not always easy to observe but one can see the results of its activities: small, round holes in leaves. While the adult is a leaf-eater, the larvae, like those of most longhorns, gnaw holes in wood. Eggs are laid in fissures in bark at the base of young trees. The larvae at first eat the bark but continue into the wood making ascending tunnels which end in holes through which they eject "sawdust". Considerable damage may occur. Larval development takes 2–3 years.

197 Longhorn-beetle
Rhagium mordax
L 10–19 mm   June–July   Common; deciduous woods

A typical longhorn-beetle despite the comparatively short antennae. It is often found sitting on tree trunks or flowers; in warm sunny weather it flies readily. The larva lives beneath the bark of various deciduous trees, especially birch. It attacks those which have been previously infested with insects and are dead or dying. The larva tunnels between the bark and the wood and makes a pupal chamber out of fragments of wood.

198 Red Poplar Leaf-beetle
Chrysomela populi
L 10–12 mm   May–Aug   Local; on aspen and sallow

Both adults and larvae of this species feed on leaves and soft plant materials. The adult hibernates in fallen leaves on the ground. In spring, females lay eggs on the under-sides of leaves of sallow and aspen. The larvae eat the entire leaves leaving only midribs and veins. If a larva is disturbed it secretes an evil-smelling fluid from apertures on either side of the body. This is a protective mechanism against predators. Pupation takes place in July–August and new adults emerge a week later.

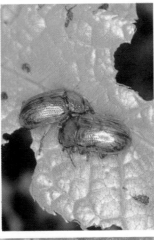

199 Leaf-beetle
Phytodecta species
L 5–7 mm   May–July   Common; deciduous woods

The light-green larvae (left) of these beetles live on leaves of various deciduous trees and bushes in particular sallow, willow, rowan, bird cherry and aspen. If they occur in large numbers they completely defoliate a bush in a very short time. The adults (right) also feed on foliage. They are usually reddish-brown or black. One common species *P. decemnotata,* which occurs on aspen in much of Britain, is distributed throughout the Northern Hemisphere.

200 Leaf-beetle
Agelastica alni
L 5–7 mm   May–July   Very rare in S. England; on alder

This leaf beetle, a rarity in Britain, has specialized on alder. The eggs are laid on the leaves which the larvae later devour, leaving only the veins. Elsewhere, mass occurrences of this little beetle completely defoliate whole trees. It tends to be more common in coastal districts.

201 Green Leaf-weevil
Phyllobius maculicornis
L 5–11 mm   May–June   Local; deciduous trees and bushes

These beetles occur in spring and early summer being very active in sunny weather. The body is covered with tiny gleaming scales which tend to fall off when the insect is handled. These scales can be green as in this species or bluish or copper-coloured. In weevils the head is always more or less elongated into a rostrum at the end of which the mouthparts are situated. The adult eats leaves of various bushes and young trees. The larvae live in soil and feed on roots.

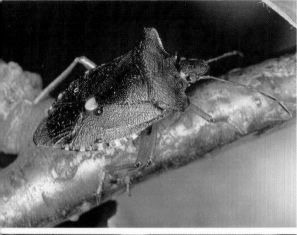

202 Forest-bug
Pentatoma rufipes
L 12–16 mm Feb–Nov Common;
deciduous woods, gardens

Found mainly on oak and alder but
also on fruit trees-especially cherries
where they damage fruit by tainting
it with a characteristic unpleasant
odour. Like most pentatomids they
are predatory and catch insect
larvae and drain their body juices.
The "forest-bugs" hibernate as
nymphs in bark fissures high up in
trees. Many of them serve as food
for birds in winter. They become
adult in July and eggs are laid on
leaves in August. Females usually
survive longer than males, often until
late autumn.

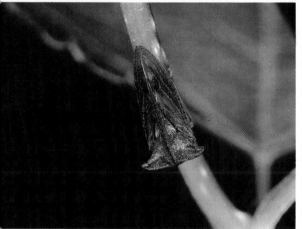

203 Tree-hopper
Centrotus cornutus
L 7–10 mm June–Aug Rare;
scrub, deciduous woods

This insect belongs to a family of
bugs, the "tree-hoppers", of mainly
tropical distribution. This is one of
two species found in Britain and both
are rare. They are characterized by
their hunch-backed form which is due
to the thoracic shield being extended
backwards and sideways, often in
grotesque shapes. It is a pest of
vineyards in southern Europe. The
nymphs do not have the horn-like
projections of the adult, but the
thorax is raised and the abdomen is
long and pointed.

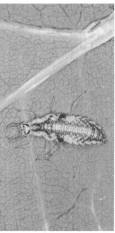

204 Lace-wing (eggs and larva)
Chrysopa species
Larva L 6–11 mm May–Aug
Common; on deciduous trees, bushes
and herbs

Eggs (left) are attached to leaves
by long stalks. The female puts a
drop of mucus on a leaf and pulls it
into a thread. When it has hardened
she places an egg on it. The larvae
(right) are voracious and useful
predators on aphids etc. In some species they pile their cast skins and
the empty skins of their prey on
their backs. This may disguise them
and deceive their predators.

205 Great Green Bush-cricket ♂
Tettigonia viridissima

L 40—55 mm May—Oct Locally common; by trees, bushes etc.

This species sings in summer sitting in bushes or trees high up. The song is exceptionally loud and can be heard 200 yards away. It consists of harsh bursts of high-pitched singing that last for up to 20 secs. After a moment's pause the sound is repeated so that the general effect is of sustained continuous song. It is produced by the male which rubs the bases of its fore-wings against each other. In Britain the species is restricted to the southern counties of England and Wales. Song diagram:

```
|0        |5        |10       |15 sec
```
time

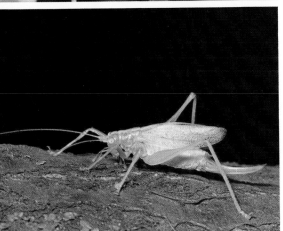

206 Oak Bush-cricket ♀
Meconema thalassinum
L 13—17 mm May—Nov Common; on deciduous trees, mainly oak.

This species is more widely distributed in Britain than the previous one. It lives mainly in the foliage of oak trees and, being nocturnally active, is difficult to find. In autumn females may be found on the tree trunks where they lay eggs in bark fissures. They may also lay in oak-apple galls (No 208). The males do not stridulate but attract females by drumming with their hind-feet on the substrate e.g. an oak leaf. Unlike other bush-crickets this species is exclusively herbivorous.

207 Plaited Door Snail
Marpessa laminata
H 14—18 mm April—Oct Common; on trunks of deciduous trees

One can often see these slim snails creeping on tree trunks (especially beech and oak) when the air is very humid. They are also met with on rocks and stone walls in the vicinity of deciduous woods. When it is dry they retire beneath the ground vegetation or under stones. They feed on fungi and rotting plant material.

208 Oak-apple gall; Gall-wasp
Biorhiza pallida
D gall, 25 mm   June–July, Oct–Nov
Common; on oak

These pale galls are common in spring on young oaks. They gradually become darker and in the summer winged males and wingless females emerge. The female lays eggs on the roots of oak and the larvae from these produce irregular, single-chambered root galls. From these a winter generation of wingless females appears which lays eggs parthenogenetically on the outmost tips of the branches. The galls are the reaction of the plants to chemical substances produced by the larvae.

209 Marble gall
Cynips quercus-folii
D gall, 15 mm: L wasp 3–4 mm
April–May, Sept–Nov   Common; on oak

These round galls on the under-sides of oak-leaves are familiar objects. If dissected in autumn, inside will be either a pupa or a tiny black wasp (right). These wasps, all females, hibernate in the gall on the ground and in spring lay eggs parthenogenetically in buds on the lower branches of oaks. These produce both males and females and after fertilization the latter lay eggs on the under-side of oak leaves and a gall forms around each.

210 Leaf galls on alder; gall-mite
Eriophyes laevis
Gall 2–3 mm; mite 0.2–0.3 mm
June–Oct   Common; on alder leaves

Galls are formed by many different kinds of insects: wasps, flies, beetles, aphids, psyllids, scale insects and moths. They can also be caused by mites, nematodes, bacteria, and fungi. The galls pictured here are caused by very small mites. These are almost worm-like, although anteriorly they bear two pairs of legs. They live on the juices of leaf tissues which they macerate with their minute but sharp jaws. When the leaves are shed in autumn, the mites crawl into winter buds or under bark to hibernate.

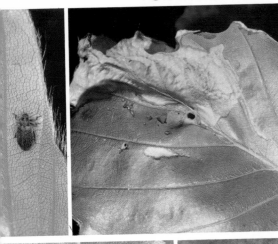

211 Leaf mines on beech, mining beetle
Rhynchaenus fagi
L 2–3 mm   May–June   Common; on beech

This mine is caused by a little jumping weevil (left). The adult beetle eats the young leaves in the spring and the females lay their eggs in the midribs. At first the larva makes a mine along the midrib and then proceeds to eat the green tissues leaving the cuticle intact. In this way a blister mine is formed which wrinkles and turns brown (right). The larva pupates within the mine in a white cocoon. In some years the foliage is extensively damaged by the activities of this insect.

212 Rolled leaves on birch: Leaf-rolling weevil
Deporaus betulae
L 3–5 mm   May–July   Common; on birch and alder

The female of this weevil is responsible for the conical leaf-rolls (left) that are common on young birches. The little black weevil first cuts a slit from one side of the leaf to the midrib and then does the same on the other side. It rolls up one side towards the middle and then the other round the first. Deep incisions in the leaf flaps keep the rolls in place. The larvae live in this shelter until they drop to the ground to pupate.

213 Robin's Pin-cushion gall on rose; Gall-wasp
Rhodites rosae
D gall 20–30 mm; L wasp 3–5 mm   May–July   Common; wild roses

This gall is caused by a little wasp which lays its eggs in a leaf bud in early spring. Instead of developing normally the bud then produces a mass of moss-like filaments surrounding a number of hard cells containing the larvae. The larvae live and the pupae hibernate in these cells. In spring, adults, mostly female and probably parthenogenetic emerge. This gall has a fascinating folklore and is also called the "Bedeguar gall" from a Persian word meaning "wind-borne".

214 Pine Hawk-moth (caterpillar)
Hyloicus pinastri
L 70–80 mm   July–Sept Very local;
on pine

Can be occasionally found on spruce
or larch but lives generally on pine.
The young caterpillar resembles a
pine needle but as it grows it acquires the characteristic body markings
seen here. Like most hawkmoths it
possesses an anal horn the function
of which is unknown. It feeds in the
day, often high up in the crown of
the tree. When full-grown it climbs
down and pupates in the ground.
The moth emerges from the reddish-
brown pupa in May–June of the
following year.

215 Pine Hawk-moth
Hyloicus pinastri
Ws 67–91 mm   May–July   Very
local; in or near pine woods

When at rest during the day on a
tree trunk, the colour of the fore-
wings blends perfectly with that of
the bark. It comes to life at dusk and
seeks fragrant flowers such as lilac,
honeysuckle and stocks. It is sometimes observed drinking sap from
exuding trees. It is not so shy as
most of its relatives. The hawk-moths
are good flyers with powerful wings
and streamlined bodies. Some species can fly at 30 mph. They can also
hover in mid-air. Caterpillar No 214.

216 Black Arches
Lymantria monacha
Ws 26–55 mm   July–Aug   Locally
common; woodland

A notorious pest of conifers on the
continent it also occurs on deciduous
trees. It is common in most of
southern England, particularly in the
New Forest. Eggs are laid in cracks
of bark and do not hatch until the
following spring. The caterpillars
feed on foliage and are full-grown by
July. Pupation occurs in a fissure of
the bark and the adult moth emerges
in late July or August. In some
European countries attempts have
been made at biological control by
spraying virus diseases from
aircraft.

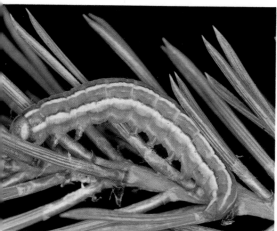

217 Pine Beauty (caterpillar)
Panolis flammea
L 30–35 mm   May–July   Common;
on pine

The adults of this species appear in
April and May and lay eggs in rows
on pine needles. The caterpillars
feed on these, which they closely
resemble in appearance, and, fully
grown by late July, enter the ground
to pupate. Winter is spent as a pupa.
In some places the species is a major
pest of conifers. In N. Germany,
where it has devastated large areas,
rather than use large amounts of
insecticides, attempts have been
made to control it biologically by
introducing a small parasitic fly.

218 Bordered White
Bupalus piniaria
Ws 25–35 mm   May–July
Common; in pine woods

This day-flying moth sometimes
causes serious damage in pine
woods. It lays eggs in rows on pine-
needles and the caterpillar, which
closely resembles a needle, eats
away the edges of each needle
leaving only the central vein intact.
Because this is not destroyed the
tree usually survives one season's
attack. When full-grown in autumn,
25–30 mm, the caterpillars lower
themselves to the ground on threads
and pupate. Moths emerge the following May.

219 Pyralid-moth
Dioryctria abietella
Ws 25–30 mm   July–Aug   Local;
coniferous woods

The plate shows a spruce cone with
the brown granular faeces of the
caterpillar of this moth. It bores its
way into the cone eating the seeds
and scales. It also attacks the
young shoots of both pine and spruce
but as a rule only when these have
been previously damaged by the
"pine shoot-moth" (No 235) or by
the "spruce bud-worm." The caterpillars descend to the ground in
autumn and pupate. The mottled
brownish moth lays its eggs on the
young shoots and cones.

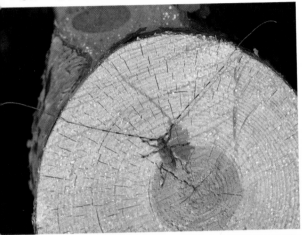

220 Timberman ♂
Acanthocinus aedilis
L 12–20 mm   April–June   Rare but
often imported; coniferous wood.

The male "timberman" has enormously long antennae; those of the
female are shorter. On warm, sunny
spring days adults walk on pine
logs and stumps in which the female
lays eggs. She gnaws a funnel-
shaped hole in the wood and then
inserts eggs with her ovipositor. The
larvae tunnel between wood and bark
and pupate in chambers made of
wood fragments. New adults appear
in autumn of the same year. This
beetle never attacks healthy wood
and is not a pest.

221 Longhorn-beetle
Monochamus sutor
L 15–24 mm   July–Aug   Doubtfully
British; conifers

The female lays eggs during late
summer in wind-felled or burnt trees
and in log piles. At first the larvae
live under the bark but they tunnel
deeper into the wood to hibernate.
In spring they eat outwards again
thus making U-shaped tunnels.
Pupation takes place in the outer
wood and the adult emerges through
a round hole. In areas of extensive
forestry this species can be a major
pest and measures have to be taken
to control it.

222 Longhorn-beetle
Rhagium inquisitor
L 8–15 mm   April–Aug   A rare
northern species; conifers

This species resembles *R. mordax*
(No 197) but is rare in Britain. Its
larvae live in stumps of conifers and
under the bark of dead trees. Even
when numerous they cause no damage and are considered useful scavengers. Before pupating they make
a chamber under the bark. The
adults usually form in autumn but
over-winter in the pupal chamber and
appear in the following spring.

223 Longhorn-beetle
Arhopalus rusticus
L 16–18 mm   July–Aug   Rare; in
Scotland, conifers

The adult of this species is found in
late summer in coniferous forests
usually near lumber or other dead
wood. Unlike most other longhorn-
beetles, which are sun-loving animals, this species becomes active
at dusk and is sometimes taken at
light. The larvae live in stumps and
trunks of pine that have previously
been attacked by other insects.

224 Longhorn-beetle
Spondylus buprestoides
L 14–24 mm   July–Aug   Common
on continent; conifers

This non-British species has powerful
jaws and will bite. The female is
much larger than the male. The
beetles often fly at dusk but can
also be found during the day. The
larvae develop in rotting wood of
both pine and spruce. They occasionally attack telegraph poles, etc.
and can sometimes be heard gnawing
the timber of newly built houses. It
is only during the first few years
after a house has been built however
that they cause damage.

225 Ant-beetle
Thanasimus formicarius
L 7–10 mm   April–Aug   Local;
conifers

This little beetle looks very like an
ant as it runs among the logs of
conifers. It is beneficial as it lives
on bark-beetles: probably the worst
enemies of the forester. In the plate
one is seen attacking a bark-beetle.
The larvae are reddish and grow to
a length of 10–15 mm. They live in
the tunnels of bark-beetles where
they prey upon the larvae, pupae and
adults.

226 Pine-weevil
Hylobius abietis
L 9–14 mm   April–Aug   Locally abundant; conifers

This unobtrusive, slow-moving weevil is a major pest of conifers. The adult attacks young (usually 2–4 year old) saplings, eating the bark so that they die. Eggs are laid under the bark of stumps or logs lying on the ground and the larvae bore into the wood. As a control young plants are treated with water-proof insecticide before being planted out and poison-impregnated bark can be used to trap it.

227 Banded Pine-weevil
Pissodes pini
L 6–9 mm   April–July   Local; conifers

This weevil is similar to the previous one but is smaller and its antennae are placed half-way along its rostrum. It occasionally attacks weakened pines but is usually confined to dead trees. The larvae make tunnels between bark and wood. The adult bores small holes in the bark of young pines causing a flow of resin and sometimes death. Several close relatives also attack conifers but these are local or rare in Britain.

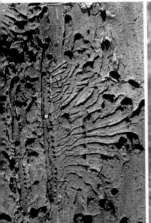

228 Bark-beetle
Ips typographus
L 4–6 mm   All year   Rare; on spruce, pine and fir

There are about 60 species of bark-beetles in Britain several of which are serious forest pests. This species, is rare here. A male and two females usually associate together and produce a vertical gallery immediately below the bark, one female working upwards and the other downwards. The male removes the debris through the entry hole and then excavates a central nuptial chamber. Eggs are laid on either side of the vertical tunnel and the larvae produce secondary tunnels at right angles to this.

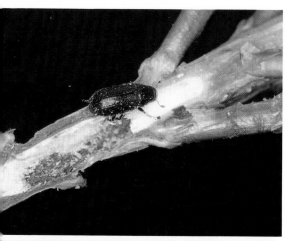

229 Bark-beetle
Blastophagus piniperda
L 4–5 mm July–Sept Local; mainly pine but also on other conifers

This is one of our most destructive bark-beetles. The adults eat into and hollow out shoots, each damaging up to three before entering the bark to hibernate. In spring they resume activity, mate and lay eggs under the rough bark of dead or dying trees or of logs. The larvae develop and pupate in the wood and the adults, emerging in July-August, ascend the trees to attack the shoots. Infested trees show large numbers of fallen shoots in autumn and their growth may be impaired.

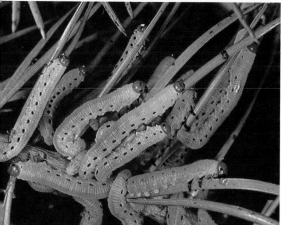

230 Pine Saw-fly (larvae)
Diprion pini
L 20–25 mm July–Sept Local; pine trees

This sawfly lays eggs in holes it makes in pine needles. They are covered with a secretion which hardens. The larvae are gregarious and their colonies defoliate branch after branch. When disturbed the larvae "lash" with the fore-part of their bodies. In autumn they enter the ground and hibernate in cocoons. Pupation occurs and the adults emerge in spring. Young needles on small trees are attacked most severely and heavy infestations seriously impair growth. Several close relatives are also pests.

231 Giant Wood-wasp ♀
Urocerus gigas
L 20–45 mm July–Sept Common; on conifers

This wood-wasp looks formidable. It is often mistaken for a large wasp but is quite harmless to man. Its larvae can cause considerable damage to conifers, although usually the adults choose fire-damaged or newly-felled trees for oviposition. The larvae tunnel the wood and when full-grown pupate. The adult emerges through a hole in the bark, the entire development taking 2–3 years. The larvae are attacked by parasitic wasps including *Rhyssa persuasoria* (see No 233) which drills deep into the wood to lay eggs near them.

232 Robber-fly
Laphria flava
L 15–20 mm May–Aug Rare;
woodland, mainly coniferous

This predator lives in glades and
clearings in woodland where it sits
"crouching" ready to dart at prey.
When a potential victim comes
within range it flies up and seizes
it in mid-air with its fore-legs. Ladybirds, bumblebees and wasps may
all be attacked. The larvae are also
predatory and live in rotten stumps
and logs. Animal food may be supp-
lemented with rotting plant material.
The related *L. marginata* is more
common in Britain's predominantly
deciduous woods.

233 Ichneumon-fly
Rhyssa persuasoria
L 20–35 mm June–July Common;
in coniferous woods

This species, our largest ichneumon,
parasitises larvae of wood-wasps
(e.g. No 231) which inhabit tunnels
in coniferous logs. The female de-
tects its hosts by means of its antennae and arching its abdomen,
forces its hair-like ovipositor, which
may be 4 cm long, through bark and
wood until it can lay eggs in the
host's tunnel. This process can
take from 10 to 30 minutes. The
Rhyssa larvae feed ectoparasitically
and ultimately kill their hosts. After
metamorphosis the young adults
gnaw their way out to the surface.

234 Money-spider
Linyphia triangularis
L 5–7 mm July–Oct Common;
meadows, woodland

The hammock-like webs of this spider are best seen on mornings in
autumn before the dew has lifted.
They are placed on grass, in bushes,
or in trees. The web is supported by
vertical threads fastened at higher
levels. When an insect flies against
one of these it falls onto the sticky
threads of the "hammock" and
becomes easy prey for the spider
waiting beneath. The male and
female live together during the
breeding period.

235 Resin "gall" of the Pine Shoot-moth
Petrova resinella
Ws 16–21 mm   June (moth)   Local;
on pine

"Pine shoot-moth" eggs are laid
behind young buds on pine. The larvae burrow into the shoots and
cause resin to flow. They hibernate
within lumps of hardened resin
(inaccurately called galls) and resume feeding in the spring. The
"galls" then enlarge. After a second
hibernation the larvae pupate.
Attacked shoots usually die and damage is greatest when leading shoots
are attacked. *Rhyaciona buoliana,* a
close relative, causes similar damage.

236 Larch adelges, gall on spruce
Adelges laricis
L 1–2 mm   May–Sept   Common;
on spruce and larch

The life-history is complex. In autumn mating occurs on spruce and
the eggs laid produce females which
lay eggs in spring. The larvae from
these feed near spruce needles
causing galls to develop. Some individuals stay and breed in the galls,
others with wings fly to larch where
breeding occurs until autumn.
Winged forms then return to spruce
and lay eggs which develop into
males and females. All broods
except the autumn one are produced parthenogenetically.

237 Bud gall on Juniper, Gall-midge larvae
Oligotrophus juniperinus
Larva: L 3–4 mm   All year
Common; on juniper

The left plate shows the very
common small galls made by this
species on the tips of juniper shoots
(N.B. Berries also present). Each gall
contains a little orange larva (right).
The midge lays eggs in young needles causing them to develop abnormally. The needles fuse together in
groups of three and become elongated. The larva hibernates inside the
gall. A traditional use of these galls
was as a cure for whooping-cough.

238 Snow-flea ♀
Boreus hyemalis
L 3–5 mm   Oct–April   Locally
common; on snow and under moss
in North

This species frequents snow when
the temperature is a few degrees
above freezing. It cannot fly but can
jump about 20 cm. The wings of the
female are much reduced whereas
those of the male are spine-like and
used for holding the female during
mating. Eggs are laid in moist soil
or moss and both larvae and adults
are moss feeders. The larvae are
produced in spring and pupation
occurs in autumn.

239 Winter-gnat
Trichocera saltator
L 4–8 mm   All year   Common; on
snow, in vegetation

Winter-gnats occur all the year round
but are most noticeable during winter
when large numbers of males form
swarms. They become active as soon
as the temperature rises a few degrees above zero. Swarming, a pre-
mating, hovering flight activity, is
impeded when the temperature increases to 17–18° C. Winter-gnats
are closely related to crane-flies but
are completely harmless. The larvae
live in soil feeding on vegetable
matter and fungi. Pupation takes
place in the ground.

240 Soldier-beetle larva
Cantharis species
L 15–23 mm   Sept–April   Common;
on ground and snow

The larvae of these beetles sometimes occur in large numbers on
thawing snow when they are forced
up to the surface by waterlogged
conditions. Otherwise they remain in
the soil or under stones. They are
predators feeding on earthworms,
snails and other small animals which
are injected with saliva containing
strong digestive enzymes to liquefy
their tissues. In spring the larvae
pupate in the soil and the first
adults appear at the time of leaf-
burst. Adult No 244.

241 Common Earwig ♀
Forficula auricularia
L 9–16 mm April–Oct Common; on ground, under stones

Earwigs are nocturnal and hide during the day under stones and in other dark places. They do not crawl into people's ears even though the fear of this is very widespread. At night earwigs can be found crawling on tree trunks and walls. During autumn the female makes a nest in which she and the male pass the winter. The eggs are laid in spring. The female is well known for caring for her small, white offspring. She keeps them together, feeds and cleans them.

242 Hairy Rove-beetle
Creophilus maxillosus
L 15–22 mm April–Oct Common; on dead animals and dung

The rove-beetles constitute the largest of our beetle families with over 800 species in Britain. The elytra are short and cover only part of the long and flexible abdomen. Beneath the elytra many species have fully developed wings which can be folded compactly. Most species are predators, some however live on rotting plant materials. This species in one of our largest and often occurs on dead animals, especially birds. It also occurs on dung where it preys on insect larvae.

243 Ground-beetle larva
Carabus species
Larva: L 15–30 mm March–Sept Common; on and in the ground

These larvae are powerfully built and are well armoured with thick sclerotised plates. They have well-developed mandibles and paired tarsal claws. Voracious predators, they feed on insect larvae, slugs and other small animals and are capable of attacking earthworms several times their own size (plate). During the day they burrow in soil and litter but at night come to the surface in search of food. They pupate in cells in the soil. Adult No 244.

244 Ground-beetle
Carabus cancellatus
L 20–27 mm April–Sept Common; woods, gardens etc.

One may sometimes hear ground-beetles rustling among dead leaves on a summer evening. They are nocturnal predators, running down their prey and devouring it on the spot. Insects, slugs, snails and earthworms can all be taken. *Carabus* species often have their elytra beautifully sculptured with ridges, tubercles and pits. This species is one of the finest. It is common on parts of the Continent but the only confirmed British record is from the S. of Ireland.

245 Ground-beetle
Carabus granulatus
L 14–23 mm April–Sept Common; marshy places

This beetle varies in colour from dark coppery-brown to dark green. It is often encountered in wet grassland but usually hides under stones and logs during the day. It becomes active at night-fall. When handled ground-beetles emit an evil-smelling liquid which, it is thought, deters larger predators, e.g. badgers and hedgehogs, from eating them. This ground-beetle can live for several years. In the first year females are not very fertile but thereafter lay about 1,000 eggs at a time.

246 Ground-beetle
Carabus nemoralis
L 22–26 mm April–May, Aug–Oct Common; gardens etc., lowland areas

Perhaps the commonest and most widespread of the 12 *Carabus* species in the British Isles. It has a rather narrow head and each elytron has three longitudinal rows of pits. The species is quite at home in urban surroundings and may sometimes be seen eating insects that have flown to street lamps and fallen down to the pavement. It is absent from the mountainous regions of Scotland but here a comparatively rare, all-black species, *C. glabratus*, sometimes occurs.

247 Ground-beetle
Feronia caerulescens
L 9–12 mm May–Oct Common;
on dry ground

Unlike most ground beetles the members of this genus are active by day and this shiny beetle is often encountered in open grassy places. The eggs are laid in spring and early summer. The larvae grow rapidly and pupate after about a month several inches down in the ground. Adults are produced in autumn but usually remain underground until spring.

248 Wood Tiger-beetle
Cicindela silvatica
L 15–19 mm April–July Locally common; on sandy ground, often in pinewoods

Tiger-beetles are fast-moving elegant animals common in dry sunny places. They are predators, flying or running down their prey with a ferocity that has gained them their English name. Their proficiency at flying distinguishes them from their nocturnal relatives, the ground beetles, and although they cannot sustain continuous flight, they readily take to wing. This species is slightly larger, but less common, than the following one.

249 Green Tiger-beetle
Cicindela campestris
L 12–16 mm April–June Common; grassland, paths etc.

This beautiful shining-green species is often seen in spring while the ground is still bare. Occasionally mating couples are found (plate). Tiger beetles have a 2-year life cycle and both adults and larvae occur together. The larva digs a vertical tunnel in the ground and, securely anchored by a hooked hump on its abdomen, lies in wait within with its head and large mandibles blocking the entrance. Ants and other small insects are seized as they come within reach.

250 Rhinocerus-beetle ♂
Oryctes nasicornis
L 25–39 mm   June–July   In heaps
of sawdust and compost

This beetle is exclusively nocturnal
and swarms around midsummer flying
noisily. It does not occur in Britain
although elsewhere in Europe it
seems to be increasing. Originally the
larvae probably lived in hollow trees
and rotting stumps but nowadays they
are found in sawdust heaps, compost
and even in cold frames. The larvae
thrive at about 25° C., development
from egg to adult takes 3–5 years
and both larvae and adults live on
rotting vegetation. Only the male
has a horn.

251 Dor-beetle
Geotrupes vernalis
L 14–20 mm   April–Sept   Local; in
dung and on decaying fungi

These beetles fly at dusk in search
of fresh animal droppings which they
locate by smell. The female digs a
hole 30–40 cm deep under the
dropping and plugs its bottom with
dung. A rabbit pellet may be used for
this purpose (plate). She lays an
egg on the plug and the larva feeds
on it. These beetles are the ''shard-
borne'' beetles of Shakespeare and,
because they are usually infested
with large brown mites are also
known as ''lousy watchmen''.
There are six species in Britain.

252 Dung-beetle
Aphodius rufipes
L 11–13 mm   May–Aug   Common;
in dung

There are over 40 species of *Aphodius* in Britain. Some are active by
day, others by night. This species
flies in darkness and flies to light,
often entering open windows.
Dung-beetles have an acute sense
of smell and seek fresh cow-pats (or
dung of horses, sheep, etc.) and
burrow inside them. They lay eggs
in the dung and the larvae have to
compete with many other insect
dung-feeding species. They pupate
in the soil and hibernate as adults.

253 Sexton- or Burying-beetle
Necrophorus investigator
L 18–22 mm   April–Sept   Common;
on dead animals

This natural undertaker has sensitive chemo-receptors and can detect the smell of carrion from a distance. Several individuals gather on a corpse and dig away the soil from underneath so that it becomes buried. After a contest for ownership a pair of beetles or a single female gain possession. A cell is made under the corpse and eggs are laid in a passage leading into this. The female feeds the young larvae with regurgitated juices from the dead animal and the larvae pupate after about seven days.

254 Carrion-beetle
Necrodes littoralis
L 15–25 mm   May–Aug   Common;
on corpses, especially on shore

This beetle specializes on corpses of larger animals and occurs in large numbers on dead cows etc. The eggs are laid directly on the body which is too large to be buried. The enzymes of the larvae break down the tissues to simpler substances which, in turn, are decomposed by bacteria. This is important in the turnover of organic substances in Nature. Carrion-beetles have an unpleasant smell and when provoked regurgitate their stomach contents which smell worse. Note the orange mites on this and the preceding species.

255 Carrion-beetle
Silpha carinata
L 11–20 mm   May–Sept   On ground

This beetle is a predator, killing and eating snails and earthworms. It does not occur in Britain but resembles the smaller British *Phosphuga atrata* which has similar habits. Another common species is the beet carrion-beetle, *Aclypea opaca* which damages crops of sugar-beet. Both adults and larvae feed on the leaves leaving irregular holes and faecal smears. The larvae resemble woodlice and the adults have a fine covering of hair. Several other types of crop can also be attacked.

256 Black-banded Spider-wasp
Anoplius fuscus
L 10–15 mm   April–Aug   Common;
sand-dunes, sandy heaths

The females of this species are
predatory on wolf-spiders. Large egg-
bearing females are preferred and
the smaller males are ignored.
Captured spiders are stung to paralyse them and are then placed in
prepared holes dug an inch or so
into sand. An egg is laid on the spider's abdomen and the wasp larva
eats the spider which survives until
its vital organs are attacked. Female
adult wasps hibernate gregariously
in holes in the ground.

257 Common Spiny-digger
Oxybelus uniglumis
L 5–6 mm   May–Aug   Common; on
sandy ground

This is our commonest digger-wasp.
It makes its nest in sandy ground and
stuffs it with flies as food for its
larvae. Most species carry prey home
with their legs but this one impales
it and carries it with its ovipositor.
This may lead to complications.
Sometimes other wasps steal the
flies as their owner enters its nest,
at other times parasitic flies choose
this moment to attach their larvae
to them.

258 Red-banded Sand-wasp
Ammophila sabulosa
L 12–20 mm   April–Sept   Common;
sandy ground

The adult hunts caterpillars of Lepidoptera as food for its larvae. It
makes a burrow in sand and covers
the entrance with a stone. When it
has found a prey it stings it, paralyses it with a nerve poison and, with
great difficulty, drags it to the
burrow. After removing the stone, it
pulls in the prey, lays an egg on it,
retires and closes the burrow. One
female may provision several
burrows. The caterpillar survives
until the larva is ready to pupate.

259 Field Digger-wasp
Mellinus arvensis
L 12–15 mm   July–Sept   Common
on sandy ground

This wasp catches flies as food for
its larvae and frequents dung patches
where they gather. It leaps onto its
prey without warning, grabs it,
paralyses it and then drags it to the
"nest" for storage. Each larva needs
6 flies to complete its development
and most of the adult's life is spent
collecting them. The "nest" is a
burrow with larval cells at the ends
of side-branches. The adult needs a
good memory for location as she has
to be able to find the opening.

260 Gooden's Nomad
Nomada goodeniana
L 8–12 mm   April–Sept   Common;
on sandy and dry grassland

The females of this "cuckoo-bee"
lay eggs in the nests of solitary bees
that live in holes in the ground.
Strictly speaking they are not parasites, as they eat only their hosts'
food, they are "social parasites".
When a female has found a nest she
watches it waiting for a suitable
moment to enter and lay her egg on
the pollen inside. The "cuckoo"
larva grows faster than the larva of
the solitary bee which usually starves
to death. Another genus of "cuckoo-
bees" parasitizes bumblebees.

261 Ruby-tail wasp
Chrysis ignita
L 6–9 mm   April–Sept   Common;
on dry and sun-exposed ground

Chrysid wasps are immediately recognized by their beautiful shining
colours. They can be seen sitting
on walls and logs in warm sunny
weather. If attacked they curl up
becoming almost spherical. They are
then protected by their extremely
thick exoskeleton. Their larvae are
parasitic on other wasps, usually
digger-wasps. The female sneaks
into the nest, lays her eggs and
retreats quickly. The eggs hatch
rapidly and the larvae live on the
food and bodies of the host larvae.

262 Carpenter-ant
Campanotus ligniperda
L 5–18 mm March–Oct On trunks and stumps of trees

Ants, like bees, wasps and many parasitic species belong to the Hymenoptera. They live in colonies consisting of fertile females (queens), males and large numbers of sterile females (workers). Nests are made in soil, under bark or in piles of pine needles and the brood are raised within them. There are no combs or cells as in colonies of bees and wasps. Many ants are omnivorous and like sweet things. This species does not occur in Britain but is common locally on the Continent.

263 Wood-ant
Formica species
L 4–11 mm March–Oct Common; in coniferous woods

Several related species build the ant hills that are conspicuous in coniferous forests. These are built of conifer needles, twigs and other plant fragments. An ant hill can be over 4 feet high and contain hundreds of thousands of individuals. In the underground parts of the nest are chambers inhabited by queens with eggs, larvae and pupae. The workers catch insects and collect honeydew from aphids to feed the larvae. They also maintain the right temperature and humidity for them. Ant ''eggs'' sold as fish food are pupae.

264 Common Black ant
Lasius niger
L 3–9 mm May–Oct Common; in soil and under stones

This ant is common in gardens where it often makes nests under paving stones and in rockeries. It is attracted to honeydew and one often sees ant paths from the nests to aphid-infested plants. On calm warm days in late summer there suddenly appear large numbers of winged individuals. These are males and females swarming. They mate on the wing high in the air. After mating the females break off their wings and form new colonies; the males die. The plate shows workers and two females.

265 Mound-ant
Lasius flavus
L 2–9 mm   March–Oct   Common;
on pasture, under stones.

This species of ant is rarely seen
because it is predominantly subter-
ranean in habit. It builds its nests in
tussocks of grass. In common with
most other types of ant, its colonies
contain numerous other species of
insects. These include root-aphids
and coccids both of which produce
secretions nutritious to the ants. A
small beetle, *Claviger testaceus,* also
occurs. This is fed by the ants in
return for the secretion it supplies.
It is blind and cannot subsist independently.

266 Red ant
Myrmica rubra
L 4–7 mm   March–Oct   Common;
in soil and under stones

Most species of ants have no sting
but spray formic acid into wounds
they inflict with their powerful mandibles. This, however, can sting.
Like other ants it likes sweet substances and this has led to an interesting association with the caterpillar of the "large blue", *Maculinea arion,* which has a gland producing a sugary exudation. The
caterpillars are captured by the ants
and taken to the nests where the
gland is "milked". In return the
caterpillar feeds on the immatures of
the ant (See No 24).

267 Ant-lion
Myrmeleon formicarius
Larva: L 10–12 mm   All year   In
uncompacted sand and earth,
mainly in pine woods

"Ant-lions" are the larvae of nocturnal dragonfly-like insects. They are
mainly tropical and none occur in
Britain. The species illustrated,
however, is found on dry heaths in
some Baltic islands. The "ant-lion"
digs a conical pit in the sand and
lies at the bottom with only its mandibles protruding. Small insects fall
into the pit, their descent being
assisted by a shower of sand pro-
jected by the larva, and are seized
and sucked dry. Larval development takes two years.

268 Grey Flesh-fly
Sarcophaga carnaria
L 6–17 mm   April–Oct   Common;
on dung, carrion etc.

This fly is attracted by the smell of
meat. The female deposits several
first-stage larvae in rapid succession
onto the meat and these commence
to feed secreting enzymes which
liquefy the food. Their spiracles are
adapted to act as floats so that the
larvae do not drown in the liquid.
Pupation is in the soil and the adults,
like those of most higher flies emerge from the puparium and, when
necessary, dig through the soil by
use of a retractable bladder on the
head.

269 Noontide-fly
Mesembrina meridiana
L 11–13 mm   May–Oct   Common;
on dung in pasture

This fly can often be seen on animal droppings, on walls, tree-trunks
and on the ground. It sucks nectar
from flowers and is frequently
observed on umbels and ivy. The
female lays single large eggs in
horse- or cattle-dung and these
hatch almost immediately. The larvae
develop within the dung.

270 Common Yellow dung-fly
Scopeuma stercorarium
L 10–12 mm   April–Oct   Common;
on dung in grassland

This yellow-haired fly occurs on
fresh cow- and horse-dung often in
such large numbers that the whole
dropping is covered. When one approaches they all fly up with a loud
buzzing noise to return shortly to the
same spot. The females lay their
eggs in the dung which serves as
food for the larvae. Pupation occurs
in the ground under the dung. The
adult fly is a predator; it catches
other flies and sucks them dry. It
also visits flowers.

271 Goat-moth, caterpillar
Cossus cossus
L 80–100 mm   All year   Common;
in deciduous wood, on ground

This caterpillar lives in various trees
e.g. poplar and ash making long
winding tunnels in the wood. It takes
2–3 years to mature. In autumn it
leaves the wood and searches for
hibernation and pupation quarters
in the ground. Normally meat-red, it
lightens just before pupating. The
smell is unpleasant and penetrating
and, although they have no olfactory
sense, birds seem to avoid this species. Pupation occurs within a cocoon
of soil and wood fragments. (Moth
No 171).

272 Summer-chafer, larva
Amphimallon solstitialis
L 20–30 mm   All year   Locally
common; sandy grassland etc.

The larvae of this and other chafers
are well known to farmers and gar-
deners as white grubs. They live on
the roots of various plants and can
be very destructive to crops and
grass. They also attack young pine
trees. The development of this
species usually takes two, sometimes
three, years. Adult No 193.

273 Click-beetle, larva
Agriotes species
L 20–30 mm   All year   Common; in
arable land and pasture.

Click-beetle larvae are predominantly herbivores. Some species live in
rotting wood but larvae of this genus
live on roots and root organs of
various plants. They are the notorious
"wireworms". In general they thrive
best in grassland but can devastate
cereal and potato crops that have
been grown on recently ploughed
land previously under grass. Potatoes
that have been attacked by wireworms are unsightly and tend to
rot quickly. Arable land older than
2–3 years is not usually susceptible
to intensive wireworm attacks.

274 Mole-cricket
Gryllotalpa gryllotalpa
L 35–46 mm   All year   Rare; in
loose earth often near water

The "mole-cricket" resembles the
mole in many ways. Its body is
covered with fine hair, it burrows
through the ground with its specially adapted fore-legs and it feeds on
worms, insects and roots. Occasionally it damages garden and cold-frame
crops. Recently it has become rare,
possibly due to man's activities such
as drainage. The insects fly on warm
summer nights and the males "sing".
Eggs are laid in holes in the ground
in spring and early summer and the
life-cycle lasts two years.

275 Wolf-spider
Lycosa lugubris
L 5–8 mm   March–Oct   Common;
on dry, sunny ground, mainly in
woodland

Wolf-spiders do not build snare-
webs but jump on their prey and
stab it with poison-claws. Enzymes
are injected and the liquefied
contents are sucked out. The plate
shows mating. The grey male transfers sperm from his pedipalps to the
genitalia of the female and she lays
eggs on a web placed on the ground.
She rolls this up, attaches it to her
spinnerets and carries it with her.
After hatching the young spiders
remain on her back for about a week.

276 Zebra-spider
Salticus scenicus
L 4–7 mm   Febr–Oct   Common; on
walls, rocks and tree trunks

This is a jumping-spider i.e. it takes
its prey by surprise jumping on it
from a considerable distance. It
regulates its jump by attaching an
anchor line to vegetation. Like the
wolf-spiders, it has four pairs of
eyes and a well-developed sense of
vision. This species is commonest in
warm sunny-places. It is active during
the day and spends the night in small
silken cells. It also hibernates and
lays its eggs in these cells.

277 Harvestman
Phalangium opilio
L 5–9 mm   All year   Common; on ground, in cellars and sheds, on bushes and trees.

Harvestmen often have long legs and lack the "waist", poison and silk of true spiders. They run after their prey which includes insects, mites and slugs. If caught by a leg, the leg is shed but does not regenerate. The second pair of legs function as antennae. Using her ovipositor the female lays eggs deep in the soil or in cracks in bark. The young hatch in spring and the adults mature by late summer. They sometimes hibernate gregariously in shelter.

278 Red Velvet-mite
Trombidium species
L 0.5–4 mm   April–Sept   Common; on the floor of deciduous woodland

Mites belong to the Arachnida the large group of jointed-legged animals containing spiders and harvestmen. Most have 3 pairs of legs as larvae and 4 pairs when adult. These large species are common in woods under logs and in leaf litter during spring. They are predatory on small insects and other mites. Their larvae are parasitic on insects etc. and may often be seen firmly attached to the legs and bodies of harvestmen.

279 Woodlouse, Common slater
Oniscus asellus
L 13–18 mm   All year   Common; in damp and shaded places

Woodlice belong to the Crustacea, the major group of aquatic arthropods. Although most species have adapted to terrestrial conditions they are very sensitive to desiccation and are found in damp surroundings e.g. among dead leaves, under stones and in damp, unheated buildings. They are active at night when the humidity is high feeding on rotting plant remains. Beneath her body the female has a brood pouch where the eggs and young are kept. The young develop slowly becoming full-grown after about two years.

280 Pill-millipede
Glomeris marginata
L 13–18 mm All year Common; on the ground, under stones, in tree stumps

The "pill-millipede" rolls into a ball when disturbed. It can be confused with the "pill-bug", *Armadillidium,* a woodlouse which, however, is grey. The "pill-millipede" is common in woodland and undisturbed grassland especially on calcareous soils. It feeds on rotting plant materials and fungal mycelia. While egg-laying the female lies on her back and letting the eggs extrude from the genital opening, passes them back from one pair of legs to the next until they reach the anus. Then each is covered in a protective coat of excrement.

281 Snake-millipede
Cylindroiulus punctatus
L 20–30 mm All year Common; in moss and ground litter, under bark

The name millipede means "thousand-footed" but no species has that many. Characteristically, most body rings (diplo-segments) bear two pairs of legs. When the animal moves each pair of legs bends after the preceding pair so that wave-like motions appear to run from front to rear. On the sides of the body are stink-glands whose secretion protects them against enemies such as birds. They eat rotting leaves, wood and some fresh vegetation and sometimes cause damage in kitchen gardens. This is the commonest species of the genus in Britain.

282 Centipede
Lithobius forficatus
L 20–30 mm All year Common; in soil, under bark and stones

Whereas millipedes appear to have two pairs of legs per body "segment" centipedes have only one. Unlike millipedes they are predators and the legs of the first segment behind the head have been modified to form poison claws. Like its relatives this centipede is a nocturnal hunter feeding on insect larvae, earthworms and other small animals. The female carries her eggs between the hind legs before depositing them in the soil. Sometimes she rolls them on the ground to give them a protective covering of soil.

283 Centipede
Brachygeophilus truncorum
L 35–45 mm   All year   Common; on moist ground, under bark and stones.

This centipede also has the first pair of legs modified into poison claws for killing earthworms and insect larvae after it has wrapped itself around them. They are also cannibals, attacking their own young. Occasionally they eat vegetable matter e.g. potatoes and fruit. These centipedes crawl backwards and forwards with almost equal facility – this is useful when crawling in and out of bark fissures and other restricted places. Some members of this group of centipedes produce a phosphorescent secretion particularly when disturbed or handled. This light can be seen from several yards.

284 Earthworm
Lumbricus terrestris
L 100–300 mm   All year   Common; in the soil of pastures and arable land, in gardens

The body of an earthworm is divided into segments. The sexually mature worm has a thickened section called the "saddle" which, during mating, secretes material which hardens to become a capsule around the eggs. They are hermaphrodites and during copulation there is a reciprocal exchange of semen. Earthworms are useful; their burrowing causes a much needed turnover of air and humus in the soil as well as draining it. They eat dead leaves and and other rotting plant material.

285 Large Black slug
Arion ater
L 100–130 mm   May–Oct Common; in woodland, on rough land etc.

This slug has the breathing hole leading into the respiratory cavity near the front of the mantle (plate). The two pairs of tentacles, on the posterior pair of which the eyes are situated, are drawn into the body if the animal is disturbed. The colour is very variable, most are black but white, grey, and brown individuals occur. These slugs eat fresh and rotten plant materials, excrement and dead animals. The eggs are laid in the soil in clusters. Hibernation occurs as the egg or small juvenile.

286 Brown Aeshna ♀
Aeshna grandis
L 67–76 mm　July–Sept　Common;
beside water, on fields and in woodland clearings

The male is brown like the female
but differs in having blue eyes and
a more slender abdomen with small
blue markings anteriorly. This dragonfly captures and eats its prey on
the wing. The prey – flies, midges,
small butterflies, etc. – are grasped
by the legs which are held out in
front of the body to form a capturing
device. The female lays eggs on submerged plants just below the water
surface. Larval development takes
2–4 years, the adult usually lives 1–2
months. Larva No 341.

287 Southern Aeshna ♂
Aeshna cyanea
L 70–78 mm　July–Sept　Common;
near water, in fields and pastures

This beautiful dragonfly is one of
the most common species in the
south of England but is absent from
Scotland. It can be found at considerable distances from water and
even finds its way into houses. In
folklore dragonflies have been ascri-
bed with various frightening and
unpleasant traits. From the human
point of view they are completely
harmless. In Europe they are known
as devil's horses (Denmark), troll-
flies (Sweden), and devil's needles
(Germany); in no case terms of
endearment.

288 Common Aeshna ♂
Aeshna juncea
L 67–76 mm　July–Sept　Common;
near lakes and ponds, in pine woods

This is a common dragonfly in northern Europe and can be met with at
woodsides and in gardens. Like
other *Aeshna* species it is an accom-
plished and swift flyer being able to
hover and dart off in any direction
with speed and agility. It can even fly
backwards! The head is dominated
by huge compound eyes which together possess some 2,800 separate
elements. It also has 3 small simple
eyes on the forehead. Because of
their sharp vision and agility, dragonflies are very hard to catch.

289 Golden-ringed dragonfly ♂
Cordulegaster boltoni
L 75–83 mm May–Sept Common;
along streams, mainly in hilly country

This large species is commoner in
central and southern Europe than
in the north and is found hawking
along streams in wooded areas
or sitting on vegetation resting or
eating prey. The female has a conspicuous ovipositor sticking out from
the end of the abdomen. With this
she deposits eggs in the sandy bottom of shallow water while she hovers just above the surface. The
large voracious larvae live in the
sludge on the bottom but leave the
water before turning into adults.

290 Club-tailed dragonfly ♂
Gomphus vulgatissimus
L 48–51 mm May–June Local and
rare; streams and woods

Females of this species are sometimes met a long way from water.
Though rare in Britain and confined
to the southern counties it is plenti-
ful on the continent and is widely
distributed from the Pyrenees in the
west to oriental Russia in the east.
The species is characterized by the
club-shaped abdomen. It always
emerges early in the morning the
larvae having crawled out of the water during the night. The metamorphosis from larva to adult is very
short; less than two hours.

291 Downy Emerald
Cordulia linaenea
L 46–50 mm May–Aug Locally
common; near water

This "hawker" dragonfly can often
be seen patrolling a shoreline. It can
hover in the air and will defend its
"territory" against intruders of the
same or related species. In mating
(plate) the male places sperm into a
receptacle near the front end of his
abdomen. He then grips the neck of
the female with his abdominal claws
and she curves her abdomen forwards
under his to receive sperm from the
receptacle. This species resembles
the somewhat later species *Soma-
tochlora metallica*, the "brilliant
emerald".

292 Four-spotted Libellula
Libellula quadrimaculata
L 39–48 mm   May–Sept   Common;
near still water

When, after 2–4 years, the larva is
full-grown, it crawls out of the water
one morning onto a reed and holds
on firmly. The skin splits down the
back and the soft wet pale dragonfly
appears – at first head downwards.
It curls upwards, grasps the skin with
its legs and draws out its abdomen.
Later it pumps air and fluids into its
wings so that they expand. After
2–5 hours its body has become
sufficiently dry to permit flight.
Adult No 293.

293 Four-spotted Libellula
Libellula quadrimaculata
L 39–48 mm   May–Aug   Common;
near still water

It may take up to two weeks after
emergence before the adult attains
full colouration and sexual maturity.
The plate shows a newly emerged
adult. The dark spots on the front
margins of the wings are characteristic for this species. It watches for
prey from a branch and darts out
whenever a suitable insect approaches. Mating lasts a few seconds after which the female lays eggs on the
water during flight. This species
may migrate in very large numbers; in
1852 it appeared in a swarm 500 m
long.

294 Broad-bodied Libellula ♂
Libellula depressa
L 39–48 mm   May–Aug   Common;
still water, woods

The broad abdomen of this species
is brownish yellow in the female and
blue in the male. This is a more
agile and faster flyer than the preceding species but otherwise has a
similar way of life. Like most other
large dragonflies it is hard to find
in rainy and overcast weather as it
hides in trees and bushes. During
oviposition the female flies low and,
repeatedly dipping her abdomen into
the water, deposits a few eggs at a
time.

295 White-faced dragon-fly
Leucorrhinia dubia
L 34–39 mm   May–July   Locally
rare in North; near bogs

This is the smallest *Leucorrhinia* in
northern Europe and the only one
in Britain. Their white foreheads
distinguish them from species of
Sympetrum. The plate shows mating
with the male uppermost. Because
of their mottled appearance these
dragonflies are difficult to detect
when at rest on a background of
vegetation. The female lays about
500 eggs, 10–15 each time she dips
her abdomen during flight into the
water. Disease parasites and
predators cause heavy mortality of
immatures.

296 (Nordic) White-faced dragon-fly ♂
Leucorrhinia rubicunda
L 37–42 mm   May–July   Common
locally on continent; fens, sphagnum
bogs

This species, which does not occur in
Britain, is difficult to distinguish
from its near relative pictured above
although it gives the impression of
being more powerfully built. The
males and, to a lesser extent, the
females commonly bask in sunshine
sitting on stones, logs and tree
stumps. As in other white-faced
dragonflies the female is always alone during oviposition. The similar
Sympetrum females, however,
commonly fly in tandem with males
when they lay eggs.

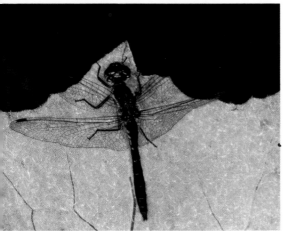

297 Black Sympetrum ♂
Sympetrum scoticum
L 31–35 mm   July–Oct   Locally
common; peaty ponds, heaths, pine
woods

Like other members of this genus
the ''black sympetrum'' rarely flies
over water but usually stays some
distance from its edge sunning itself
on dry ground. The females and
young males have conspicuous yellow
and brown markings on the body. In
favourable weather mass occurrences
can arise, when almost every stem
has an empty larval skin on it and
hundreds of dragonflies fly up with
each step one takes.

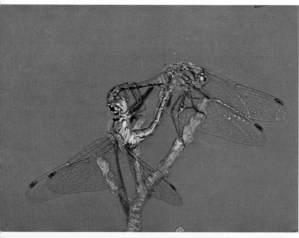

298 Ruddy Sympetrum
Sympetrum sanguineum
L 33–38 mm   June–Sept   Local;
weedy ponds

The plate shows a mating pair, male
uppermost. Mating is as with No 291;
the male first transfers the sperm
from his posteriorly-situated genitalia to the receptacle of the acces-
sory organs below the second abdominal segment. This organ has
claspers which grip the end of the
female's abdomen as the sperm is,
in turn, transferred to her. Mating couples can fly almost as well
as single individuals. This migratory species occurs only in the southern half of Britain.

299 Ruddy Sympetrum ♀
Sympetrum sanguineum
L 33–38 mm   June–Sept   Local;
weedy ponds

Oviposition in *Sympetrum* species
often takes place when the males
and females are still joined together.
Repeatedly the animals drop to the
water surface to release a few eggs
at a time. The plate shows a female
that has just been released by the
male. Two eggs are hanging from
the abdomen. Members of this genus
and of the related genera *Libellula*,
Leucorrhinia and *Orthetrum* are
known as ''darters''.

300 Yellow-winged Sympetrum
Sympetrum flaveolum
L 34–38 mm   July–Aug   Rare;
small weedy ponds

This species is easily recognized by
the large yellow spots on the wings
(in exceptional cases these can be
absent). Like its close relatives it
tends to rest on or near the ground
at some distance from water. At
rest the wings are usually held in a
forwards-downwards position. This
dragonfly is very difficult to
approach.

301 Demoiselle Agrion ♂
Agrion virgo
L 45–48 mm   May–Aug   Common;
by tree-lined streams and rivers

This damselfly occurs in large numbers along unpolluted streams in sunny glades. Damselflies are slow flyers and their fluttering flight is reminiscent of that of butterflies. Occasionally a male can be seen courting a sitting female, standing in front of her with his wings flapping rapidly. Males are always more numerous than females and when a female takes to wing she is often pursued by a dozen or so males following in a long undulating line until one succeeds in mating with her. Female No 302.

302 Demoiselle Agrion ♀
Agrion virgo
L 45–48 mm   May–Aug   Common;
by tree-lined streams and rivers

The body of the female has a greenish coppery lustre and the wings are usually brown. In older individuals the wings become violet. At egg-laying she crawls backwards down a stem protruding through the water and lays eggs inside it just below water level. Sometimes she descends further becoming completely submerged. The larvae live among vegetation in the water probably for two years. Adult life lasts, however, only two weeks. Pollution of streams is causing this species to become scarce.

303 Banded Agrion ♀
Agrion splendens
L 45–48 mm   May–Aug   Locally common; by streams

This damselfly is commonest along large streams and rivers where the bottom is muddy. It frequents rather more open places than the preceding species but is only found in the southern half of the country. The sexes are markedly different in appearance. The female (plate) is shimmering green with transparent green wings but the male is blue with a broad blue (brown when immature) band crossing each wing.

304 Large Red damselfly ♂
Pyrrhosoma nymphula
L 33–38 mm   April–Aug   Common;
ponds, bog-pools etc.

This is the larger of two species of
red-bodied damselfly found in Britain. It can be recognised by the
brown stripes on the thorax. A poor
flyer it takes sessile prey such as
midges and aphids. During oviposition the sexes remain attached and
descend into the water together. The
female deposits eggs in holes bored
in the under-water stems of plants.
While submerged these insects are
vulnerable to attack by fish etc. In
the air, birds and spiders are their
chief enemies.

305 Red-eyed damselfly ♂
Erythromma najas
L 35–38 mm   May–Aug   Locally
common in Midlands and South;
lakes and scrub
This species is usually seen flying
just above water or sitting on floating
leaves. It is a better flyer than other
damselflies. The female has yellow-
green eyes and green and black
markings. Oviposition takes place up
to 2 feet below the surface of the
water by females still *in copula,* the
eggs being often placed in stems of
the yellow water lily. The submerged
adults may breathe air trapped in the
fine hair covering the body.

306 Common Ischnura ♀
Ischnura elegans
L 30–33 mm   May–Sept   Common;
near stagnant water and rivers
The male of this species has bright-
blue and black markings. The abdomen of the female is usually blue
with black markings while the colour
of the thorax varies greatly in different individuals: green, blue, violet,
orange, pink and brown being found.
Frequently the entire range of colours
is found in individuals from a small
area. Some of the colour variation
is due to differences in age.

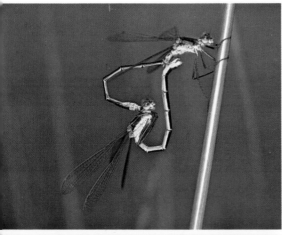

307 Green Lestes
Lestes sponsa
L 35–38 mm   July–Sept   Common;
near stagnant water

This is one of the commonest damselflies in late summer. It does not
fly fast and can easily be taken by
hand. The female is coppery-green
and brown whereas the male has
light blue markings. Damselflies
spend the night on vegetation near
water. In the morning they are covered with dew and almost immobile
with cold. Later they warm up in the
sunshine and fly again, mating and
laying eggs. The plate shows a male
(above) and a female mating.

308 Variable Coenagrion ♀
Coenagrion pulchellum
L 32–38 mm   May–Aug   Common;
near stagnant water

This is the commonest of several
rather similar species of damselfly.
The males may be distinguished by
the colour patterns of the abdomen
and by the shape of the abdominal
claws. The females, however, are
more difficult to separate especially
as colour variants occur. In this species one type of female is blue and
black as in the plate, the other is
green and black as in the plate
accompanying No 309.

309 Variable Coenagrion
Coenagrion pulchellum
L 32–38 mm   May–Aug   Common;
near stagnant water

The plate shows the process of
oviposition. The male stands upright
holding the neck of the female firmly
with his abdominal claws. She curves
her abdomen beneath a floating leaf
and inserts her eggs into it one by
one as she slowly circles round. The
abdomen gradually dips deeper during this process so that a spiral
of eggs is formed on the underside
of the leaf. Larva No 342.

310 Mayfly ♂
Cloeon dipterum
L 5–11 mm May–Aug Common;
near still water

This small mayfly lacks the hind pair
of wings. The male, apart from having
the usual pair of eyes, has another
pair of stalked ones which may aid
in the detection of females in poor
light. Mayflies and their larvae are
important food items for fish, especially salmonids. In fly-fishing, baits
are used that resemble mayflies. On
calm summer evenings the males
often swarm over lakes and the surface can be covered with empty
larval skins. Larvae Nos 343–344.

311 Mayfly ♀
Ephemera danica
L 16–24 mm May–Sept Common;
around small clear ponds etc. with a
sandy bottom

This is one of the largest mayflies
in N. Europe. The larva lives about
two years in the water usually buried
in the sandy bottom. It feeds on
organic particles. On metamorphosis,
the mayfly first appears as a winged
but immature "subimago". It moults
again to produce the adult. These
stages are the "duns" and "spin-
ners" of anglers. Adults live only a
short time, they eat no food and
their guts contain only air.

312 Alderfly
Sialis lutaria
L 10–16 mm May–June Common;
around all kinds of fresh water.

This slow insect is found in spring
sitting or crawling on vegetation and
stones beside water. The females lay
eggs close together on leaves of
various water plants. Such clusters
can be two inches long and contain
up to 2000 eggs. The larvae are
predators and live about two years
on the bottom, often at considerable
depths. They crawl onto land in
spring and pupate in cocoon-like
holes in the ground at the edge of
the water. The adults live only a few
days.

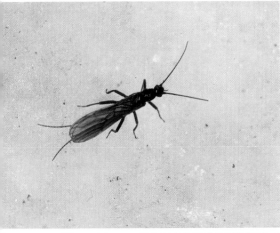

313 Stonefly
Capnia species
L 5–10 mm   Feb–May   Local; beside streams and rivers

Most stoneflies emerge in early spring after the larvae have spent about a year in running water. The adults sometimes appear on snow near streams (plate). They are slow poor flyers as a rule and some have reduced wings and are devoid of flight. Like mayflies they are important as food for salmon and trout. They hold their wings folded horizontally over the body when at rest.

314 Caddisfly
Phryganea grandis
L 15–21 mm   June–Aug   Beside standing fresh water and brackish water.

Caddisflies date back some 200 million years. Most species are active in twilight or at night and fly by day only if disturbed. They come to light readily. The adults look rather like moths but are hairy not scaled. The larvae are aquatic and build cases around themselves using pieces of plants, gravel, shells etc. fastened together with silk. Pupation is in the case and the pupa ascends to the surface before the adult emerges. Eggs are laid in mucilaginous masses often around stems.

315 China Marks moth
Nymphula nymphaeata
Ws 20–30 mm   June–Aug   Common; in vegetation around lakes, river-banks

Like most of its relatives this moth is active at night but flies by day when disturbed. Eggs are laid on leaves of water lilies and other aquatic plants and the larvae construct water-filled shelters from fragments of leaf. It feeds on the leaves and increases the size of this shelter as necessary. Pupation occurs within silken cocoons and, on emergence, the adult rises to the surface dry and ready to fly.

316 Leaf-beetle
Donacia marginata
L 8–11 mm   May–July   Local; on plants in and near water

All 15 of the British species of *Donacia* are beautifully coloured with a metallic lustre. They occur on plants in or near water and being sun-lovers are active and fly readily in warm weather. Their eggs are laid on the under-side of leaves of water plants and their aquatic larvae feed on the plants and obtain air from them direct by inserting tail-spines bearing breathing pores into their air-filled tissues.

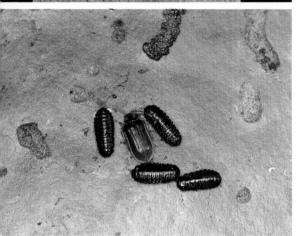

317 Water-lily leaf-beetle
Galerucella nymphaeae
L 6–8 mm   May–Aug   Locally common; on leaves of water lilies

This beetle often occurs in vast numbers on colonies of water lilies on which both adults and larvae feed. The plate shows 4 black larvae and one adult as well as a number of eaten patches. Although the leaf can sometimes be almost entirely eaten, the thin underside is usually left intact so that water cannot come through onto the leaf.

318 Rove-beetle
Stenus species
L 2–7 mm   April–Nov   Common; by and on water

Most British species occur near fresh water and some can walk on it. Occasionally one shoots across the surface like a tiny speed-boat. Species with this capacity secrete a liquid which lowers the surface tension behind them so that they are drawn forwards by the excess tension in front. They are predatory and with their large heads and protruding eyes are reminiscent of small dragon-fly larvae. Also their third pair of jaws is modified for seizing prey (cf. dragon-fly larva's mask, No 341).

319 Slender ground-hopper
Tetrix subulata
L 9–14 mm All year Common in
south; in water meadows, swamps
etc.

Ground-hoppers are easily recognized by their pronotum which is tre-
mendously elongated to cover the
the entire abdomen. They hibernate
as adults (or full-grown nymphs)
whereas most grasshoppers spend
the winter as eggs. Ground-hoppers
are most easily seen in spring, but
even then because of their colour
they are very well concealed. They
are completely mute. The species
illustrated occurs predominantly
near water and is able to swim,
both on as well as under the water.

320 Large Marsh Grass-hopper
Stethophyma grossum
L 21–37 mm May–Nov Local in
south; water meadows, fens etc.

This grass-hopper is restricted to wet
places, usually where there is open
water. It is recognizable by the black
and yellow tibiae and red femora
of the hind legs. It becomes active
in warm sunny weather and can fly
up to 10 m. Eggs are laid in autumn
at the bases of grass tussocks in
batches of about 12. They hatch in
May–June of the following year.
The song is different from that of all
other grass-hoppers consisting of a
series of ticks. Song diagram:

IIIIIIII

time
0   5   10   15 sec

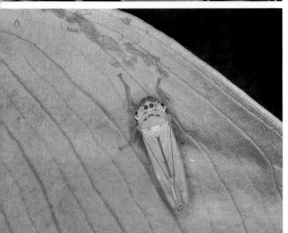

321 Leaf-hopper
Cicadella viridis
L 6–9 mm May–July Common;
water meadows, lake-shores

This is one of the largest and commonest leaf-hoppers in northern Europe.
It can move about rapidly in vegetation by jumping with its specially
adapted hind-legs. It is also very
well camouflaged. The mouthparts
are in the form of slender hair-like
stylets resting in a grooved sheath.
They are adapted for piercing and
sucking plant tissues being able to
penetrate deeply to extract sap
from the vascular regions.

322 Crane-fly
Erioptera species
L 5–7 mm   May–Sept   Common;
lake shores, edges of swamps,
stream banks

These graceful, long-legged flies
often swarm over water and lake-
side vegetation. They look like large
mosquitoes but are unable to bite
and probably only feed on nectar
from flowers. Their larvae live in
moist humus-rich soils most of them
eating roots and rotting vegetation.
Others, however are predatory on
small insects etc. There are many
species in Britain.

323 Non-biting midge ♂
Chironomidae
L 1–13 mm   March–Oct   Common;
near water

There are 400 or so known species of
Chironomidae in Britain and probably
many more await discovery. They
vary greatly in colour, grey, green,
yellow or black, and in size. ''Swarming'' often occurs on calm summer
evenings when multitudes of males
hover and dance together over land-
marks such as bushes and trees.
Females enter these and select
mates. In folk-lore swarms are
construed as signs of impending fine
weather. The bushy antennae of the
male may perceive vibrations. Larva
No 347.

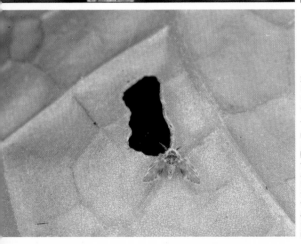

324 Hairy moth-fly
Psychodidae
L 1–4 mm   All year   Common; in
vegetation near water, indoors

These tiny hairy flies resemble small
moths. Their larvae occur at the edges of lakes etc. both above and
below water. They feed on rotting
vegetation and animal excrement and
are common on sewage farms. The
European species are harmless but
their tropical relatives, the sand-
flies, are notorious as blood-suckers
and transmitters of disease. The fly
in the plate did not cause the hole
in the leaf.

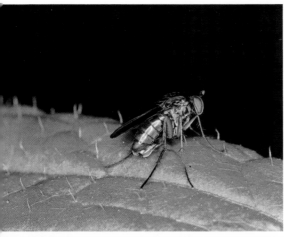

325 Dolichopodid fly
Dolichopus species
L 3–6 mm   May–Sept   Common;
near water, wet places e.g. in woodland

These flies have unusually long
coxae and when resting hold their
somewhat laterally compressed
bodies at an angle to the surface.
The genitalia of the males are very
large and are folded beneath the
abdomen. Most species have a beautiful metallic green or blue colouration. Both adults and larvae are
predatory; the latter live in water or
moist earth.

326 Raft-spider
Dolomedes fimbriatus
L 10–20 mm   April–Oct   Local;
lakes, fens, marshes

This species, one of Europe's largest
spiders, can run on water and, when
alarmed, dive beneath its surface.
Normally it rests on a floating leaf
and surveys the water surface waiting for prey to fall onto it. The female carries her egg-sac beneath
her body attached to her spinnerets
with silken strands and further supported by the chelicerae. Occasionally she dips the sac into the water
to prevent her eggs from dying of
desiccation.

327 Amber snail
Succinea putris
L 16–22 mm   April–Nov   Common;
in vegetation near water

This snail is often found crawling on
vegetation in wet places, especially
after rain. During dry periods it
stays in the immediate proximity of
water and can even be found on floating vegetation. It lives for 2–3 years.
The ''amber snail'' is the intermediate host of a small platyhelminth
worm, *Leucochloridium macrosto-
mum*. The tentacles of the parasi-
tized snail swell enormously so that
they come to resemble fat grubs.
This attracts birds which are the
final hosts of the worm.

328 Water scavenger-beetle
Hydrobius fuscipes
L 6–8 mm   All year   Common; in
stagnant water

Many beetles live in freshwater. This
species and its relatives are detritus-
and plant-feeders occurring among
submerged plants in ponds etc. They
are relatively poorly adapted for
aquatic life compared to the diving
beetles (Nos 329–30) and are mostly
poor swimmers. Their maxillary palps
are large and antenna-like; the true
antennae are tucked beneath the
head. Submerged beetles carry an
air supply forming a layer below the
body which is replenished periodically at the surface. Eggs are laid in
air-filled cocoons attached to water
plants.

329 Great diving-beetle
Dytiscus marginalis
L 30–35 mm   All year   Common; in
stagnant water

These diving beetles are proficient
swimmers with long oar-like hind-
legs. They swim with simultaneous
strokes of both hind-legs. Now and
then they swim to the surface, stick
out their rear end and fill their
tracheae with air. They also renew
their reserve supply stored under the
elytra. Both adults and larvae (No
340) are ferocious hunters and attack
fish fry, tadpoles, etc. The male has
suction pads on its fore-legs with
which it grips the female during mating. The beetles live for several
years.

330 Diving-beetle
Hydroporus species
L 2–5 mm   All year   Common; in
water

There are over 100 species of diving
beetles in Britain and almost one-
third of these belong to the genus
Hydroporus. These frequent almost
all kinds of water, ranging from clear
pools and running streams to peat
bogs and stagnant pools. Some can
live in brackish water or in the high
salinities of rock pools or salt flats.
Diving beetles are usually good
flyers and fly long distances from
one body of water to another. They
hibernate as adults, larvae or eggs
depending on the particular species.

331 Whirligig-beetle
Gyrinus species
L 3–7 mm   March–Oct   Common;
on still or slow-running water

These beetles skim around on the surface of the water propelling themselves by means of their middle- and hind-legs which are flattened and fringed with hair. They dive down into the water when disturbed or about to lay eggs. Their bodies are beautifully streamlined and adapted to an aquatic way of life. The eyes are divided into an upper aerial part and a lower aquatic part. The larvae, like the adults, are predators living among submerged vegetation. Ten species of *Gyrinus* are recognized in Britain.

332 Common pondskater
Gerris lacustris
L 8–10 mm   April–Nov   Common;
on the surface of ponds and lakes

Pondskaters are long-legged bugs that run on the surface of water. This is possible because of the surface tension of the water but they must continually keep their legs oiled with a water-repellent substance from glands on the head. Pond skaters live on midges etc. which fall on the water. The fore-legs are small and only used for grasping the prey. The adults hibernate on land in vegetation or under stones. A close relative is the "water-cricket" (*Velia* species) which occurs in streams.

333 Water-measurer
Hydrometra species
L 9–12 mm   May–Oct   Common;
on the shores of lakes and ponds

Water-measurers are inactive and stride slowly around on water or in vegetation near the shore. They feed on water-fleas, mosquito-larvae etc. which they impale on their pointed rostrums. They also take insects struggling on the surface. Unlike pondskaters, their fore-legs are not modified for predation. After mating on the water, the female attaches her large spool-shaped eggs to plants or stones above the waterline. The larvae live at least three years. (Bluish animals in the plate are spring-tails, No 351).

334 Backswimmer
Notonecta species
L 15–17 mm   All year   Common;
in standing water

These insects swim upside-down.
Their hind-legs are modified for
swimming being paddle-like and
fringed with hairs. Air is stored in
two furrows beneath the body. They
are alert predators and promptly
dive onto small animals falling in
the water. They can bite humans
painfully. *Notonecta* flies, especially
in sunshine, and sometimes mista-
kes shiny surfaces like cars for water. Its eggs are laid in aquatic
plants.

335 Water-boatman
Sigara species
L 4–10 mm   All year   Common;
usually in still water

There are more than 30 species of
water-boatmen in Britain. They live
on the bottom and among submerged
plants in ponds etc. They are herbivores and feed on algae and decomposing plant remains they stir up
from the bottom with their shovel-
like fore-legs. From time to time
they surface to renew their supply
of oxygen. They fly well. The adults
hibernate and eggs are laid in spring.
They are sometimes referred to as
lesser water-boatmen by those who
call backswimmers water-boatmen.

336 Saucer-bug
Ilyocoris cimicoides
L 12–16 mm   All year   Common in
south; in standing water

Like back-swimmers this bug can
bite man painfully. It is a predator
and its fore-legs are modified as
prehensile claws. It feeds on
aquatic insects, tadpoles and even
fish fry. It is a fast swimmer and, like
the diving-beetles, having an air
supply between wings and abdomen
can stay submerged a long time without having to come up for air. It
hibernates in the adult stage. The
female lays eggs in spring in rows on
stalks of submerged plants.

337 Waterscorpion
Nepa cineria
L 18–22 mm   All year   Common; in shallow water

This predatory bug usually stays near the bank often among water plants. It is a poor swimmer and drowns easily if it cannot reach the surface with its caudal breathing spine. The fore-legs are adapted for capturing various small aquatic animals including fish fry. Mating and egg-laying take place in spring and the animals become adult after 5 moults. A close relative is the "water stick-insect" *Ranatra linearis,* a slender bug with an extraordinarily long caudal spine. This occurs mainly in the south in Britain.

338 Water-spider
Argyroneta aquatica
L 8–15 mm   All year   Common; in brackish water

The only known truly aquatic spider. When submerged it has a silvery sheen due to air trapped in its water-repellent pubescence. Both males and females spin dome-shaped webs attached to submerged vegetation and filled with air brought from the surface. Inside these they feed, mate, rest and hibernate. After mating the female spins a new dome and fastens the eggs to its roof. Winter domes are extra dense and completely closed. Water spiders catch their prey under water but emerge to eat them.

339 Water-mite
Hydrachna geographica
L 5–8 mm   All year   In standing water

There are about 250 known species of water mites in Britain. They vary in colour, red, green and blue being common and are mostly small (L 1–2 mm). This species, our largest, occurs in shallow ponds etc. especially those with ample vegetation. The adults and nymphs of water-mites are predatory but the larvae are parasitic on many small water animals. No 332 shows *Gerris* infested with two of them. These mites have piercing, stylet-like mouthparts. Their eggs are laid on water plants or on stones.

340 Great diving-beetle, larva
Dytiscus marginalis
L 60–70 mm   May–Sept   Common;
in stagnant water

Diving-beetle larvae are voracious predators. They feed on tadpoles and insect larvae and can cause serious losses in fish ponds by eating the fry. They seize their prey with their long sickle-shaped mandibles (plate) and pass enzymes down a fine canal which penetrates these organs. These digest the tissues of the victim and the resulting liquid is sucked in through the mandibular canals. They pupate in earth at the water's edge and the whole life-cycle takes only a few weeks. (Adult No 329.)

341 Brown Aeshna, larva
Aeshna grandis
L 40–45   All year   Common; in standing water

These larvae are powerfully built predators. Their third set of jaws is adapted to form a claw-bearing "mask" which, at rest, is folded beneath the head. When a suitable prey approaches this can be shot out rapidly to seize it. The larvae usually crawl about but they can move forwards rapidly by sucking water into the hind-gut and expelling it forcibly – a form of jet propulsion. The wall of the hind-gut also functions as a gill. Larval development takes 3 years. Adult No 286.

342 Variable Coenagrion, larva
Coenagrion pulchellum
L 12–16 mm   All year   Common; in standing water

The larvae of damselflies have three long appendages on the abdomen which function as external gills and caudal fins. They usually crawl in submerged vegetation where, because of their colour, they are inconspicuous. They are predators eating small crustaceans, worms, etc. They do not take such a long time to develop as the dragonfly larvae (No 341). In this genus the larval period varies from 10 months to about 2 years. Other damselflies take only 2 months. Adult Nos 308–9.

343 Mayfly, larva
Cloeon dipterum
L 7–10 mm   All year   Common; in weedy ponds etc.

These larvae are found among vegetation in ponds etc. often in large numbers. They live on rotting plant remains and on algae growing on plants and stones. When disturbed they swim off quickly, flapping up and down with a serpentine movement. All but the youngest have tracheal gills on either side of their abdomen. They are important food for many aquatic predators, including fish. The larvae live for up to a year but the adults have a very short life-span. Adult No 310.

344 Mayfly, larva
Heptagenia sulphurea
L 9–14 mm   All year   Common; in running water

This is an example of an animal that has visibly adapted to its environment. It lives in streams and rivers often where the current is quite strong. It has to hold itself flush against the substrate or under stones if it is not to be swept away. Its body is therefore very flat and presents a minimum of resistance. It feeds on algae growing on stones. The downstream movement inevitable in running waters is compensated by the adults flying predominantly upstream.

345 Stonefly, larva
Nemoura species
L 5–10 mm   All year   Common; mainly in running water

Stonefly larvae can be recognized by their long antennae and paired cerci. Their growth occurs mainly during the colder parts of the year. As a group they are extremely adaptable to various environmental conditions and therefore occur in all kinds of water: some *Nemoura* species even occur in small ponds. They feed on algae and rotting plant materials. Stoneflies are adversely affected by human interference with their environment such as water impoundment, pollution etc. Adult No 313.

346 Caddisfly, larva
Limnephilus flavicornis
L 19–23 mm   June–Sept   Common;
in standing water

Caddisfly larvae commonly build
cases around themselves consisting
of plant materials, sand grains,
small stones and even snail shells.
They are usually cylindrical and open
at both ends. The larva passes its
entire development inside this shelter, sticking its head and part of its
thorax out when it feeds. It eats
rotting plant materials. Some larvae
are predatory and do not build cases,
some spin webs in the water to catch
plankton. The latter are commonly
found at the ouflows of lakes.

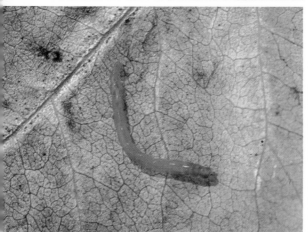

347 Chironomid midge larva,
    ''bloodworm''
Chironomus species
L 15–20 mm   March–Oct   Common;
in lakes

Chironomid midge larvae usually live
in sludge at the bottom of lakes etc.
where they eat organic particles.
Some species live in tubes with only
their heads protruding. There is little oxygen in the sludge but the
larvae take it up through their body-
wall and some (see plate) have the
red pigment haemoglobin to enable
them to store it. For fish feeding on
bottom deposits, chironomids are a
major food item. Adult No 323.

348 Midge larva
Chaoborus crystallinus
L 10–12 mm   March–July
Common; in standing water

This larva can float horizontally in
the water. It has two kidney-shaped
air bladders, one fore and one aft
(plate) connected to the gut, and by
regulating these it can rise or fall.
It is a predator and uses its prehensile antennae to catch water-fleas
etc. Its abdomen bears a tuft of
bristles used for swimming. These
larvae often occur in dense shoals
and the adults, which resemble
chironomids, swarm in large numbers.

349 Black-fly larva
Simulium species
L 8–10 mm   March–June   Common;
in running water

Black-fly larvae attach to rocks to
avoid being washed away by strong
currents. Bristles on the head filter
organic particles from the water to
serve as food. The larvae survive in
rapids and as such water is well-
aerated they have no problems of
oxygen supply. Full-grown larvae
construct funnel-shaped cases and
pupate within them. The adults rise
to the surface in bubbles of air,
males emerging first and collecting
in vast flying swarms over water
waiting for females. Adult No 406.

350 Mosquito, larva and pupa
Aedes cantans
Larva: L 7–8 mm   Dec–April
Common in south; in standing water

Mosquito larvae hang from the water
surface by their breathing tubes.
They have a group of bristles round
their mouths which are used for filter feeding. Long hairs elsewhere
serve both as sensory organs and
for steering. When disturbed,
these larvae descend to the bottom
with a characteristic wriggling
motion. Larvae hatch in autumn and
pupate in spring. The pupa breathes
through horn-like projections on the
thorax and hangs from the surface
with them. Adult No 405.

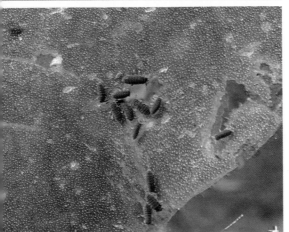

351 Spring-tail
Podura aquatica
L about 1 mm   All year   Common;
on standing water

Spring-tails or Collembola are primitive forms thought by some experts
not to be insects at all. They are
wingless but can move rapidly by
jumping with the aid of a fork-shaped
organ at the tip of the abdomen. This
at rest folds below the body but it
can be straightened rapidly to kick
the body forwards. Collembola
occur in many environments and are
prominent in leaf litter where they
break up plant material. Some deep
soil species cannot jump. This species lives on the surface of pools etc.

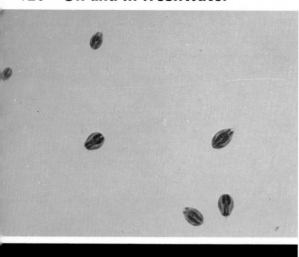

352 Ostracod
Notodromas monacha
L about 1 mm   All year   Common; in ponds, bog pools, lake shores

These crustaceans are ubiquitous in bottom deposits and among aquatic plants. Like bivalve molluscs (No 363) their bodies are enclosed in paired shells. They swim actively using their antennae and legs as oars. The species shown swims up-side-down just below the surface in sunny weather. Other species frequent the bottom and feed on detritus particles they stir up with their antennae. Some are plant eaters and some are predatory. Eggs are de-posited on water plants and may hatch after 20 years desiccation.

353 Copepod
Cyclops species
L 1–3 mm   All year   Common; in standing water

These crustaceans lack shells. *Cyclops* is a freshwater form but other genera have marine species. The antennae are used for swimming, giving the animals a peculiar jerky motion and the male also uses his for holding the female during mating. Some copepods feed on small par-ticles, others are blood-sucking parasites of fish. The female usually carries her eggs in paired sacs attached to her abdomen (plate). Some species however lay eggs free-ly in the water.

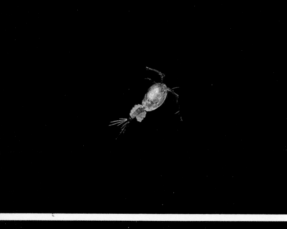

354 Water-flea, Cladocera
Daphnia species
L 2–5 mm   All year   Common; in standing water

Water-fleas resemble copepods in having strong branched antennae used for swimming. They live on plankton strained from the water by the bristles on their legs. Most females have a brood-pouch beneath the dorsal plate in which eggs are placed and the young develop. Parthenogenesis is usual but sexual reproduction also occurs. Sometimes these animals become so abundant that the water looks like red soup. They are important as food for higher animals including fish.

355 Water-louse
Asellus aquaticus
L 15–20 mm   All year   Common;
in standing and brackish water

This isopod is related to the wood-lice although it lives in ponds and streams. It occurs among shore vegetation and on the bottom feeding on decomposing vegetable matter. It also occurs in brackish water. In some places a cave-dwelling form exists that has completely lost all pigmentation. A relative, *Saduria entemon* is a well-known relict that still exists in northern lakes and seas where it was left behind by the receding glaciers of the Ice Age.

356 Freshwater shrimp
Gammarus lacustris
L 20–25 mm   All year   Common;
usually in standing water

This crustacean, an amphipod, thrives best among the stones on the bottom but it can swim rapidly to safety if it encounters danger. Like most freshwater shrimps, it swims on its side. It feeds on dead plants and animals but will also attack smaller living animals. The female has a brood pouch under the thorax where the eggs and larvae are carried. The illustrated species occurs mainly in lakes; a close relative *G. pulex* inhabits running water. These animals constitute an important fish food.

357 Aquatic worm
Tubifex species
L 30–80 mm   All year   Common; on the bottoms of lakes and rivers

Tubifex lives in sediment on the bottom within a tubular case. That in the plate has left its tube. Normally the head is in the tube and the tail is exposed but if the animal is disturbed it withdraws completely. It thrives best in running water but can survive in polluted and oxygen-deficient waters by extending further out of its tube and waving about. Oxygen is absorbed through the skin and through the hind-gut. These worms are sold as food for aquarium fish.

358 Planarian
Polycelis species
L 10–12 mm   All year   Common;
in standing water, streams, springs

Flatworms, or turbellarians occur in
large numbers in various aquatic
habitats from springs and clear, fast-
running mountain streams to sluggish
rivers and ponds. At first sight they
look like small leeches but they lack
the annulations on the body. They are
covered with tiny, moving hairs,
cilia, by whose action they are able
to move around. They are both predators and scavengers. These animals
are extremely primitive: the absence
of an anal opening is an indication of
this. They are hermaphrodites.

359 Worm-leech
Erpobdella octoculata
L 50–60 mm   All year   Common;
running or still water

Leeches are related to earthworms
but their bodies are flattened and
lack the microscopic bristles of the
latter. They have suction pads at
both ends of the body, a large one at
the rear and a smaller one at the
front. This species is carnivorous
and attacks small worms, insect
larvae etc. Other species are blood-
suckers mainly on fish and water-
fowl. One European species can draw
blood from Man. This is *Hirudo me-
dicinalis*, the "medical leech", which
was formerly used to remove blood
from fevered patients.

360 Leech
Glossiphonia complanata
L 10–30 mm   All year   Common; in
lakes and rivers

Leeches of this genus vary in colour
from greenish to brown. They have
2–3 pairs of dorsally-placed eyes.
This species cannot swim but holds
itself on to submerged vegetation.
It sucks the contents of snails and
insect larvae for food and will
crawl inside empty snail shells for
shelter. Leeches are hermaphrodite
and this one and its relations care
for their young in that the "parents"
carry the newly hatched young around
on their backs.

361 Great ram's-horn-snail
Planorbarius corneus
D 20–30 mm   All year   Common;
in vegetation-rich lakes and rivers

This snail lives in hard, nutrient-rich
waters in England and Wales. It
is a herbivore feeding on algae and
decomposing plant tissues and is
sometimes used in aquaria to keep
the glass clean. Despite the fact
that it is a lung-breathing snail it
does not have to surface often because the blood contains haemoglobin and can store oxygen. It also ab-
sorbs some oxygen through the skin.
There are several smaller species
of flat-coiled snails in various types
of freshwater habitats.

362 Great pond-snail
Lymnaea stagnalis
H 50–60 mm   All year   Frequent; in
vegetation-rich lakes and rivers,
also in brackish water

This is a pulmonate snail i.e. the
mantle cavity functions as a lung.
Now and again it must crawl to the
surface and exchange the air in this
cavity. Like other lung-breathing
snails it is hermaphrodite. At mating
two snails exchange spermatophores
with each other. They lay eggs on
submerged vetetation in long egg-
capsules each containing several
hundred eggs. This species is also
used for keeping aquaria free from
green algae. The genus is represented in Britain by several species
with a similar way of life.

363 Swan-mussel
Anodonta anatina
L 70–80 mm   All year   Common; in
lakes and rivers

"Swan-mussels" burrow into the
bottom sediments and are not easily
found. Usually one finds the empty
shells washed up on the bank. They
feed on micro-organisms filtered
from the water. The eggs are stored
for a time on the gills and then the
young larvae attach themselves to
and encyst in the skin of fish. After
living parasitically for a time they
fall off and burrow in the substrate.
The "freshwater-mussel", *A. cygnea*
and the "painter-mussel", *Unio
pictorum,* also occur in Britain.

364 Spindle-shell
Neptunea antiqua
H 130–140 mm   All year   Common;
on soft bottom in deep water

This is one of our largest shells and is often washed up on northern shores. Similar to the common whelk, *Buccinum undatum,* it differs in having pronounced wavy ridges running longitudinally along the whorls. Both species are omnivorous; *Buccinum* sometimes attacks fish caught in nets. They breathe through gills placed in the mantle cavity and there is an elongate siphon. The egg capsules (left plate) are common on the shore. They may contain over 1000 eggs of which 10 may survive– at the expense of the others.

365 Pelican's-foot shell
Aporrhais pes-pelecani
H 40–50 mm   All year   Common;
on soft bottom 15–40 m deep

The opening of this turreted shell is drawn out into a flat lip with finger-like projections resembling a bird's foot. (That shown is damaged). This species is common and empty shells are often found on the shore. They are especially abundant where fishing nets are hung out to dry as they easily become entangled with them. The animal burrows into muddy gravel and feeds on small algae and other organic matter it takes in with its elongated rostrum. It has pointed tentacles and stalked eyes.

366 Tower-shell
Turritella communis
H 40–60 mm   All year   Common;
on soft bottom at 15–40 m deep

This very common species inhabits offshore muddy banks. It seems to prefer a muddy gravel substrate and burrows into its uppermost layers. Densities of more than 100 per square metre can occur. It feeds on planktonic particles that are carried into the mantle cavity in the respiratory current and are filtered off by the gills. Empty shells are washed up on the shore and may often be found to be occupied by hermit crabs.

367 Flat periwinkle
Littorina littoralis
H 10–13 mm   All year   Common;
on shores, especially on wracks

This flat-topped sea-shore snail is
extremely variable in colour. It is
usually yellow, but brown, orange,
red and green individuals occur.
It feeds on seaweed on which it also
lays its flat egg clusters. The newly
hatched young look like diminutive
versions of the adults. Large numbers
of these snails are washed up by
storm waters around the shores.

368 Common or Edible periwinkle
Littorina littorea
H 20–25 mm   All year   Common; on
rocks in the tidal zone

This widely-distributed snail and the
related ''rough periwinkle'', *L.
saxatilis*, occur in large numbers on
rocky shores. *L. littorea* feeds by
scraping minute algae from off the
rocks and inhabits a lower part of the
mid-tide zone than *L. saxatilis*.
The shell varies in colour from dark-
grey to reddish-brown with darker
rings (plate). The animal lays its
eggs, three at a time, in small capsules. The young live freely in the water for several weeks and then become sedentary. Winkles are traditional
sea foods.

369 Common limpet
Patella vulgata
L 50–70 mm   All year   Common;
on rocks

This limpet occurs all round our
coast and attaches itself to rocks in
the tidal zone. Each limpet has its
place on a rock, its ''home'' where
its shell fits perfectly. When disturbed, or when the sea is rough it
holds on so tightly that it can hardly
be dislodged. When the tide is in it
moves over the surrounding area
browsing on algae. A gap between
its shell and the rock allows water
to flow over the gills. It returns to
its home before low tide.

370 Common mussel
Mytilus edulis
L 60–100 mm   All year   Common;
on stones etc. in shallow water

Mussels are bivalves i.e. their shell
consists of two halves joined together by an elastic tendon which
acts as a hinge. They can be opened
and closed by muscles. The mussel
attaches itself to rocks with a number of tough horny threads (byssus)
which are secreted by the foot.
These are broken when the animal
moves but can be formed again.
The mussel forms dense beds in the
middle and lower tide zones. It is
edible and is marketed either fresh
or preserved.

371 Flat oyster
Ostrea edulis
L 100–150 mm   All year   Common;
on hard substrates

Oysters live in creeks and estuaries
mainly round our southern coasts and
occur in dense beds which may be
cultivated for commercial production.
They are hemaphrodite, each
individual alternating between
male and female. The eggs, up to 1
million per individual, are kept in the
mantle cavity for 7–10 days. The
newly hatched larvae swim freely for
8–14 days before settling down. Individuals can live 30 years. The
oyster is cemented to the bottom by
its shell, it has no foot and is unable
to move.

372 Clam or Great scallop
Pecten maximus
L 150–175 mm   All year   Common;
on gravel bottom 30–60 m deep

A common offshore mollusc around
our coast. It varies in colour from
yellow to brown or reddish. The lower valve is convex while the upper
one (plate) is flat. The former has
been put to various uses e.g. as an
ashtray. Around the edge of the
mantle is a row of tentacles and
between them numerous black shining eyes. It can swim forwards
(hinge hindermost) by flapping its
valves and if frightened can make a
sudden escape movement in the
opposite direction. Clams are deli-
cious when grilled with bacon.

373 Common Edible cockle
Cerastoderma edule
L 10–50 mm   All year   Common; on sandy bottom in shallow water

The cockle is a common inhabitant of the middle and lower shore around British coasts. It burrows into sand leaving its short siphons sticking up into the water. The cilia on the gills cause water to enter through one siphon and to pass out through the other and planktonic particles are sieved off and passed to the mouth. Cockles can crawl short distances by use of their large foot which is also used for digging into the sand. They are edible and are marketed in large numbers.

374 Sand-gaper
Mya arenaria
L 90–110 mm   All year   Common; on sandy bottoms in shallow water

A widely distributed species that can enter estuaries and survive fairly low salinities. It lives buried in the sand with its long siphons projecting into the water. (The two small holes in the sand can often be seen at low tide). The respiratory water stream contains organic particles (plankton) which are filtered off by the gills and eaten. The sand-gaper is edible and is considered a delicacy especial-ly in America where it is called the "soft-shelled clam".

375 Iceland cyprina
Arctica islandica
L 80–130 mm   All year   Common; on soft bottoms 5–100 m deep

This is a North Atlantic species that is generally distributed around the British coasts. The shell of the young animal is light and becomes darker with age. Frequently shells become rust-coloured due to iron precipita-tion (plate). It lies buried in mud or muddy sand with its siphons sticking up into the water. It is edible but rather tough. In America it is minced and served as "chowder".

376 Baltic tellin
Macoma baltica
L 10–20 mm   All year   Common;
on soft bottom, muddy gravel, in
estuaries

Tellins are burrowing bivalves. They
have long separate siphons and,
tolerant of low salinities, often penetrate into estuaries. This species is
abundant in the Baltic. It occurs in
fairly deep water, up to 50 m, and
is an important food for fish. Shells
washed up on the beach are often
perforated: the smallest holes are
made by a boring sponge, *Cliona ce-
lata*, the larger ones by worms of the
genus *Polydora*. The latter can penetrate limestone. Tellin shells can be
pink, yellow or white.

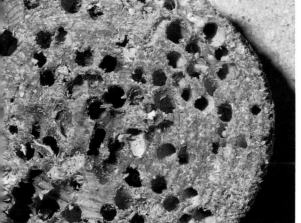

377 Timber bored by the Ship-
       worm
Teredo navalis
L (shell) 7–8 mm; body 100–200 mm
All year   Common; in wood under
water

This is a bivalve with a reduced shell
modified into a boring device. The
body is elongate and naked. Ship-
worms make broad tunnels in piles
and other submerged wood and eat
wood fragments which are digested
in the specialised gut. Proteins are
obtained from plankton taken in with
the respiratory current and sieved
off in the gills. The tunnels are lined
with lime secreted by glands on the
siphons. In the days of wooden
ships these animals were a menace;
they still cause damage to untreated
wooden piles.

378 Common star-fish
Asterias rubens
L 100–450 mm   All year   Common;
at depths down to 200 m

Star-fish belong to the Echinodermata. These animals possess characteristic organs known as "tube-
feet" which are often sucker-like
and used for locomotion. The "common star-fish" is a carnivore and
feeds mainly on bivalve molluscs. In
feeding it straddles the bivalve and
holding it with its tube-feet, exerts
pressure until the valves separate.
The stomach is then inserted and
the mollusc is digested *in situ*.
Star-fish are unisexual and fertilization is external. The larvae are
planktonic.

379 Heart-shaped urchin
Brissopsis lyrifera
L up to 70 mm   All year   Local; on clay bottoms 30–100 m deep

This species occurs in suitable places off the West, North and East coasts of Britain. It is gregarious and sometimes present in great numbers. Empty shells are washed onto the beach. In life the animals are covered with spines used for digging and defence. Brush-like tube-feet gather sand mixed with food and this is pushed into the mouth by a series of "buccal" tube-feet. Like star-fish these echinoderms are unisexual and fertilisation of the eggs occurs in the water.

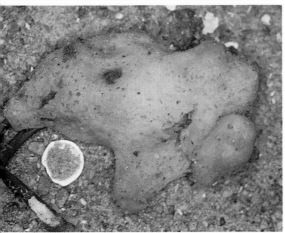

380 Bread-crumb sponge
Halichondria panicea
L up to 150 mm   All year   Common; on stones and seaweed in shallow water.

Sponges are primitive animals consisting of masses of weakly differentiated cells. They are intermediate between cell colonies and true multicellular animals. There is a siliceous or calcareous skeleton built of particles secreted by the living cells and the whole body is permeated by canals which open into pores on the surface and are lined with flagellated cells. These set up water currents which carry oxygen and food to the cells. Sponges are sessile and immobile. The species illustrated is commonly washed up on the shore.

381 Sea anemone
Sagartia troglodytes
L 20–30 mm   All year   Common; on stones, shells and plants 5–40 m deep

Several species of sea anemone are common around Britain, this one particularly in the North. They belong to the group Coelenterata whose members possess stinging cells which, when stimulated throw out poisonous threads which stick to and paralyse or kill their prey. The cylindrical body is firmly attached to the substrate by its base and at the upper end is a crown of tentacles surrounding the mouth. The tentacles move continuously and are provided with stinging cells. Large anemones in deep water catch fish and large crustaceans.

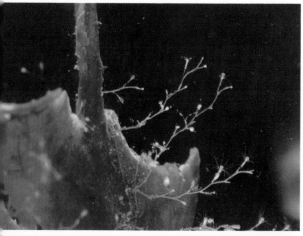

382 Sea fir
Laomedea flexuosa
L 30–40 mm All year Common;
on seaweed, stones etc. in shallow
water

This little coelenterate forms a
covering of tiny branched stalks on
seaweed etc. At the ends of the branches there are small cups each
containing a single polyp. The polyps
catch food particles with their tentacles. Reproduction is both asexual
through budding and sexual by the
formation of tiny bell-shaped
medusae that are released from the
polyp and float around in the water.
In the stomachs of these medusae
gonads are formed which give rise
to the eggs and sperm.

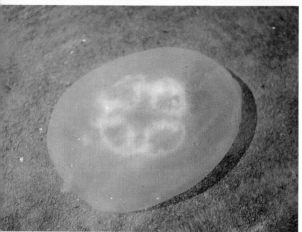

383 Common jellyfish
Aurelia aurita
D 100–400 mm June–Sept
Common; pelagic

Jellyfishes represent a line of
coelenterates where the medusal
condition has become foremost. Like
all coelenterates they possess stinging cells on their tentacles used for
paralysing prey. They are, however,
too weak to penetrate human skin.
Reproduction is complicated. The
egg gives rise to a tiny larva that
develops into a polyp. This divides
so that it looks like a pile of plates.
One after another these plates are
released and become independent
medusae. These then grow and develop into adult jellyfish.

384 Jellyfish
Cyanea capillata
D 100–500 mm July–Oct Locally
common; pelagic

Waves and currents bring these
stinging jellyfishes to our shores.
The stings are painful and swimmers
badly affected may show cramp and
have breathing difficulties. Less
seriously, they may entangle fishing
lines and cause irritation to fisher-
mens' fingers. The tentacles are
long and hang down or trail behind
entangling various animals which
are carried paralysed to the ventral
mouth. The stomach has several
lobes and numerous canals. Jellyfish
both drift passively with the current
and swim actively by contracting
their bell-shaped bodies.

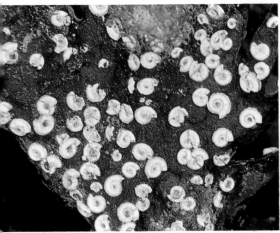

385 Bristle-worm
Spirorbis borealis
L 2–3 mm   All year   Common;
on seaweed, shells, rocks

These small spiral calcareous tubes
are made by tiny bristle-bearing
worms. They are a common feature of
British shores and those of northern
Europe. When under water they will
be seen to have fine greenish threads
hanging out of the opening. These
are the tentacles which sieve food
particles out of the water. The eggs
are laid in the tubes and hatch there
to form larvae. These disperse in the
water but after a few hours settle
and start to form new tubes.

386 Bristle-worm
Pomatoceros triqueter
L tube about 60 mm, worm up to 50
mm   All year   Common; on shells
and rocks

In this species the tube is irregular
and triangular in cross-section. The
worm has glands on the front of the
body which secrete a chalky slime
from which the tube is made. The
opening is raised slightly above the
substrate. The front end of the worm
bears red and white tentacles which
catch the small organisms on which
it feeds. The tentacles can be
retracted into the tube and the opening can be closed. This is necessary
in the tidal zone to prevent desiccation.

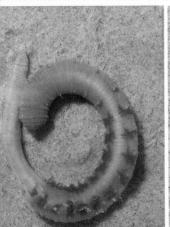

387 Common lugworm
Arenicola marina
L 100–200 mm   All year   Common;
on sand in shallow water

This worm is abundant in tidal flats
and in sandy bays but is inconspicuous because it lives in U-shaped
burrows. It possesses 13 pairs of red
gills (left) over which a forwardly
directed current of water passes. It
ingests large quantities of sandy
material, digests any matter of nutritive value, and eliminates the re-
mainder as casts (right). The lug-
worm is used by anglers for bait.
They either dig it up or make it
come to the surface by stamping on
the sand.

388 Dinoflagellate
Noctiluca miliaris
D 0.5 mm Aug–Oct Common;
pelagic

Boating on a late summer or autumn evening off the coast of Scotland or elsewhere with cool, clear seawater one may notice a momentary green phosphorescence in the wake of the boat. This is caused by a dinoflagellate. The plate shows a group of them taken from the sea on an August night. The animal is a spherical unicell with an indentation serving as a mouth. Near this is a tentacle and a short flagellum. These animals reproduce by binary fission or by spore formation.

389 Bryozoan, Sea-mat
Electra pilosa
L 0.5–1 mm All year Common; on seaweed, rocks etc. to 80 m deep

These small sessile animals occur as honey-comb-like colonies encrusting seaweed. Each cell contains a tiny individual which, when submerged, sticks out its whorl of tentacles into the water. Bryozoans are hermaphrodite and both sexual and asexual reproduction (budding) occurs. Colonies are formed by budding. This group of animals has existed for over 400 million years and many extinct forms are preserved as fossils. There are several very common and widespread species around our coasts.

390 Acorn-barnacle
Balanus balanoides
D 10–20 mm All year Common; on rocks, boulders, shells

Bathers often scratch themselves on the shells of these familiar barnacles. The larvae are free-swimming but later attach themselves by their antennae to rocks etc. and, remaining in this upside-down position, surround themselves with the hard plates of the shell. When under water the plates open up and the long hairy legs are protruded (plate). These make repeated grasping motions in the water and catch small animals and other plankton for food. They also produce a respiratory current over the gills.

391 Shore-crab
Carcinus maenas
Width of shell 25–100 mm   All year
Common; shallow water

Our commonest species of crab, this species is found all round the coast of Britain and is particularly abundant around fisheries where it feeds on fish waste. Its green colour and the five sharp teeth on either side of the shell behind the eyes enable one to distinguish it from the "edible crab", *Cancer pagurus,* which is larger, brown, and has blunt lobes at the sides of the shell. Also the hind legs of the shore-crab are flattened and adapted for swimming.

392 Shrimp
Crangon vulgaris
L 40–80 mm   All year   Common; shallow water with sandy bottom

Shrimps are among the best known animals of the shore or, boiled, of the table. When alive they are adept at camouflaging themselves, changing colour to merge in perfectly with their background (plate). They dig into the sand leaving only their eyes and antennae uncovered. If threatened, they dart off and take cover. Shrimps are omnivorous eating algae, small crustaceans and worms. They have claws to grasp their prey.

393 Prawn
Leander species
L 30–60 mm   All year   Common; among stones and seaweed in shallow water.

Prawns tend to remain in concealment under stones and seaweed. They feed on seaweed and dead animals. The winter is spent in the deeper and more protected offshore waters but in spring they return to the warmer water of the shore-line and rock pools. When cooked, unlike shrimps, they turn bright pink. There are several species of prawns in British waters, one the common prawn, *L. serratus,* varies greatly in abundance from one year to another.

394 Sand-hopper
Gammarus species
L 15–25 mm   All year   Common; in shoreline vegetation

Sand-hoppers belong to a group of crustaceans known as Amphipoda. These animals are all flattened from side to side. They swim on their sides, often in pairs, the females swimming with the larger males on their backs. This precedes mating which occurs when the female moults. The eggs are kept in a brood pouch on the underside of the female's body. Sand-hoppers live on rotting seaweed etc. and may be seen swimming away when a knot of drifting weed is lifted out of the water.

395 Isopod
Idothea baltica
L 15–30 mm   All year   Frequent; in shoreline vegetation

The Isopoda, represented on land by wood-lice, are primarily aquatic. This species which seems not uncommon around British coasts is an inhabitant of the lower shore-line. It is very variable in colour, it can be yellow, brown, green, red or almost black, and in size. The variation is partly genetically determined and partly dependent on diet. The one on the left was fed on brown algae while that on the right lived on red ones. In isopods the body is flattened dorso-ventrally.

396 Sand-hopper
Talitrus saltator
L 12–16 mm   April–Nov   Common; on sandy shores

This species is common on the upper part of the shore where it occurs in masses under pieces of drift-wood or under heaps of seaweed which afford it cool and moist conditions during the day. If one lifts such debris these amphipods start jumping about vigorously. At night they leave this shelter and move freely on the beach feeding on washed-up weed. A close relative, with similar habits, but much darker in colour is *Orchestia gammarella*.

397 Kelp-fly
Coelopa pilipes
L 5–10 mm   All year   Common; on fore-shore

Kelp-flies occur on wet seaweed that has been washed up on the shore. They can be present in great numbers, especially when autumn storms have thrown up large deposits of kelp onto the beach. They lay eggs in the heaps of weed and the larvae live within and help to de-compose them. Sometimes the flies enter houses and shops and become a nuisance. They seem particularly liable to enter dry cleaners possibly because they are attracted to trichloroethylene and similar solvents used by such businesses.

398 Muscid fly
Fucellia species
L 5–6 mm   All year   Common; near and on stranded seaweed

Large numbers of flies frequent piles of seaweed lying on our shores and fly up when disturbed. Amongst these members of the genus *Fucellia* are very common. The plate shows a male and female mating. The male is uppermost and is the smaller. Their larvae are predacious and in N. America have been shown to prey on the eggs of a marine fish which are laid in the sand above high water mark.

399 Ground-beetle
Broscus cephalotes
L 16–23 mm   April–Oct   Common; in sandy places, often seashores

This ground-beetle is often found underneath drift wood and other flotsam washed up on the shore. It digs tunnels in the sand under such objects where it lurks in wait of prey. When alarmed it freezes in a special posture feigning death (right). The predatory larva (left) is found in similar circumstances. Both the larva and the adult hibernate.

400 Horse-fly ♀
Tabanus species
L 18–25 mm   May–Sept   Common;
near water, rough land, scrub

The females of these large, rapidly
flying flies have dagger-like mouth-
parts which they use to pierce the
skin of animals to obtain blood.
Sometimes they bite humans. The
males are inoffensive, have reduced
mouthparts and suck nectar. Eggs
are laid in grass or on vegetation
overhanging water and the larvae
live in damp soil or are aquatic. They
prey on small animals. After hibernation the larvae pupate in spring.
Several species are common in
Britain: that shown, *T. bovinus* has a
restricted distribution.

401 Tabanid fly ♀
Chrysops relictus
L 8–11 mm   May–Sept   Common;
damp woods and pasture

Members of this genus are characterized by the black wing patches
and by the golden-green iridescent
eyes. They hold their wings out when
at rest (plate). They fly silently and
often alight un-noticed. Their bites
are painful however and on warm
sunny days they can be most troublesome. As in all tabanids, only the
females suck blood. Eggs are laid on
plants over-hanging water and the
larvae live in water and moist soil,
where they live on vegetable detritus.

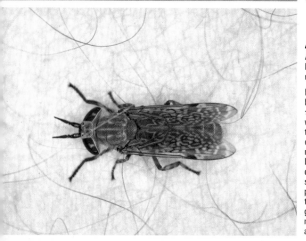

402 Cleg ♀
Haematopota pluvialis
L 8–13 mm   May–Sept   Common;
mainly near water

In these grey tabanids the wings are
held over the abdomen (plate). These
are most active in warm humid
weather and bite humans and domestic animals indiscriminately. Again,
only the female sucks blood, the
male feeding on nectar. The larvae
live in moist soil. In Britain, tabanids
cause annoyance and the larger
species may cause cattle to run in
panic (gad) and hurt themselves. In
the tropics they also transmit pathogenic trypanosomes, bacteria and
nematodes from one animal to
another.

403 Muscid fly
Hydrotaea species
L 5–7 mm   April–Oct   Common;
woods, grazings etc.

Several members of this genus are
common throughout Britain. One of
the best-known is the "head-fly",
H. irritans, which causes great annoyance to man and animals by continually buzzing round the head and
other exposed parts of the body. They
do not bite but suck sweat and can be
particularly troublesome to cattle
with open wounds. When working
energetically in the country it is
sometimes almost impossible to
avoid swallowing these flies. The
eggs are laid in animal manure and
the larvae develop within it.

404 Biting house-fly
Stomoxys calcitrans
L 5–8 mm   June–Oct   Common;
farms, stables etc.

This species looks like a "house-fly"
(No 412) except that it has long
forwardly-directed mouthparts
which it uses to penetrate the skin
and suck blood of man and animals.
Unlike mosquitoes and tabanids
both males and females suck blood.
They fly low and bite their victim's
legs; the irritation does not persist.
In some places this fly acts as a
vector of pathogenic nematodes,
tapeworms, trypanosomes and viruses. The larvae live in manure containing large amounts of straw or
hay.

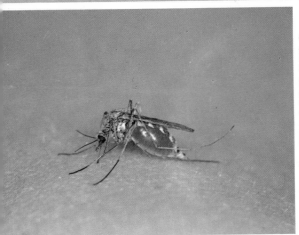

405 Mosquito ♀
Aedes species
L 4–10 mm   June–Aug   Common;
near water, in woods, by coast

Only female mosquitoes bite. Their
saliva contains an anticoagulant
which causes allergic irritation and
swelling in some people. If the fly
is allowed to engorge (plate) the
poison is partially removed with the
blood meal and after-effects are
lessened. Some of the 14 species of
Aedes in Britain are generally
distributed e.g. *A. punctatus*, whereas others e.g. *A. detritus* are confined to coastal regions. Mosquitoes
are notorious vectors of disease; *A.
aegypti* transmits yellow fever in
the tropics. The larvae live in still
water (No 350).

406 Black-fly
Simulium reptans
L 2–6 mm   April–Sept   Common;
near running water

The sucking mouthparts of the black-fly have saw-like edges which wound the skin of man and animals. These flies pester anglers and cause irritation to animals by biting them within the nostrils, ears etc. Occasionally when numerous they cause deaths among farm animals. They are worst in late summer and early autumn. The female lays eggs in sticky masses on water plants, sometimes having to dive under the water to do so. The larvae (No 349) are attached to rocks and plants.

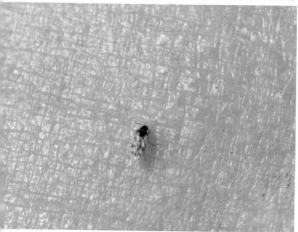

407 Biting-midge ♀
Culicoides species
L 1–2 mm   May–Sept   Common;
woods, pastures, moorland

These midges crawl on man and animals biting and drawing blood with their microscopically small mouthparts. They crawl into hair, under clothes and on all exposed surfaces and cause acute irritation. In some parts, e.g. the W. Highlands of Scotland they can be a scourge. Their presence in conjunction with that of the kilt is said to have given rise to the "highland fling". Only females bite; some species attack other insects. Eggs are laid in ponds, lakes or wet ground rich in organic matter.

408 Sheep-tick
Ixodes ricinus
L 1–11 mm   April–Nov   Common;
on rough pasture

This mite sucks blood from birds and mammals including man. The life-cycle takes three years. In summer, eggs are laid in grass and, next year, the larvae (left) climb plants, attach to hosts and feed. They moult and the nymphs behave similarly the year after producing adults. These feed in the third year, the females engorge (right) and lay several hundred eggs. Ticks cause irritation and act as vectors for diseases of man and animals. To remove one cover it with oil; it will then relax its hold.

409 Crab-louse
Phthirus pubis
L 1–2 mm   All year   Not uncommon;
on man

Two species of lice infest man. Both
suck blood and spend all their lives
on the host. They fix their eggs
(nits) to hairs (left) and have claws
on their legs enabling them to hang
on. The "crab-louse" (right) infests
the pubic region. The other, *Pediculus humanus*, has two races; one
affects the body, the other the head.
These are associated with poor hygiene and long unkempt hair. *P.
humanus* transmits diseases including the deadly typhus fever. Lice
can be controlled by insecticidal
lotions and shampoos.

410 Biting-louse
Lipeurus maculosus
L 2–3 mm   All year   Common; on
pheasants

Biting lice differ from sucking lice,
such as No 409 in having their mouthparts adapted for chewing rather than
for piercing. Many species affect
birds where they feed on feathers,
skin, dermal secretions and blood.
They are very host-specific and it is
estimated that every type of bird
has 4–5 species of lice. Some species affect mammals only. Sheep and
cattle may be heavily infested. The
plate shows, left to right, a male,
two females and a nymph.

411 Bird-flea
Ceratophyllus gallinae
L 2–3 mm   All year   Common; on
birds and in nests

Adult fleas lack wings but possess
jumping legs. They suck blood from
birds and mammals but are not very
host specific. This species affects
many birds including domestic hens.
It also bites man. It frequents nests
where its larvae feed on debris. The
pupae overwinter. The "human
flea", *Pulex irritans*, is now, due to
cleaner drier houses, a comparative
rarity. Of the world's 800 or so species several are important vectors
of disease e.g. the tropical *Xenopsylla cheopsis* is the principal transmitter of bubonic plague from rodents
to man.

412 House-fly
Musca domestica
L 7–8 mm   April–Nov   Abundant;
houses, refuse, manure

This species breeds in excrement and decaying vegetable matter and is particularly prevalent in refuse heaps. From here, laden with bacteria, it enters houses and walks over and regurgitates its stomach contents on food. Over 3.5 million bacteria may be present on a heavily contaminated fly and in one sub-tropical country infant mortality in a village was halved when the flies were controlled. The larvae are leg-less maggots and usually take about three weeks to mature. The adult hibernates in shelter.

413 Cluster-fly
Pollenia rudis
L 5–9 mm   All year   Common; in houses, on flowers, soil etc.

"Cluster-flies" occur throughout the year. In winter they are often present in large numbers in the upper parts of houses where their sluggish behaviour makes them conspicuous. In spring they go outside and suck nectar from the early flowers. They can be recognised by the way they hold their wings (plate) and by the "pollinosity" of the abdomen which produces a shifting pattern when viewed from different angles. The larvae parasitise earthworms.

414 Blow-fly, Bluebottle
Calliphora erythrocephala
L 9–15 mm   March–Oct   Common; on meat and fish

"Bluebottles" are well-known household pests and their loud buzzing noise is characteristic. In nature they are attracted by the smell of carrion and lay their eggs on corpses. The larvae (maggots) occur in masses and, excreting enzymes onto the meat, rapidly liquefy and consume it. They enter houses through open windows etc. and lay eggs on any exposed meat or fish present. On the farm they lay in wounds caused by *Lucilia* (No 51) on sheep and their larvae increase the damage.

415 Small Fruit-fly
Drosophila species
L 2–3 mm   All year   Common;
in and out of doors

These small flies collect on over-
ripe fruit. Some species are attracted
by the aroma of wine and vinegar
and are called "vinegar-flies".
They live on yeast cells and mould.
One species *D. melanogaster,* is
universally used as an experimental
animal in genetical research. The
larvae are studied in the final stages
before pupation. The reason for their
popularity is that the cells of the
salivary glands have giant-sized
chromosomes.

416 Mushroom-flies
Sciara species
L 1–2 mm   All year   Common; on
potted plants, in glasshouses

These tiny gnats are commonly
seen flying around potted plants
both in the house and in greenhouses. In general they are completely
innocuous but the larvae occasionally damage the root hairs of small
plants and can be destructive to cultivated mushrooms. Certain species
of this genus are well known for
migrating gregariously in the form of
dense columns of larvae several
layers wide. These so-called army-
worms are on the move in search of
suitable pupation conditions.

417 Bacon-beetle
Dermestes lardarius
L 7–9 mm   All year   Common; on
dried animal produce

This beetle attacks dried meat in
warehouses and larders. It also
infests untanned leather and skins.
Outdoors it occurs in bee-hives and
on dried-out corpses. It hibernates as
the adult. Eggs are laid on food-
stuffs in summer, the larva matures
in about two months and then burrows into soft material (e.g. wood)
for pupation. The pupa lies in the
tunnel with the shed skin of the last
larval instar as a protective sheath.

418 Dermestid beetle
Attagenus gloriosae
L 5–8 mm   All year   Imported
occasionally

This species is occasionally imported
into Britain in stored produce. The
larva eats furs and stuffed animals
and will attack dried insects. Recently it has been reported as a pest
on carpets. The larva (left) has bristles on its rear end and these are
raised in defence. The adult does no
damage as it feeds on pollen. *A.
pellio*, all-black with two white spots
on its elytra is a British species with
similar habits.

419 House Longhorn
Hylotrupes bajulus
L 10–25 mm   July–Aug   Rare; in
soft-wood house timbers

In parts of the Continent this beetle
causes serious damage to building
timbers. The adult is crepuscular,
short-lived and rarely seen but the
larva lives for 3–5 years gnawing
tunnels in the wood. There are no
external signs of damage and struc-
tural failure may be the first indica-
tion of infestation. The species occurs locally in Britain. House timbers should be impregnated with insecticide to prevent such damage.

420 Longhorn-beetle
Callidium violaceum
L 10–20 mm   May–Aug   Local; in
fir fence posts, house timbers

This species is of local occurrence
in Britain. In Scandinavia it is more
common and is often found around
saw-mills. If a house is built with
wood containing larvae, beetles can
appear at any time up to 3 years
later. The larvae live between the
bark and wood and are by no means
so dangerous as those of the "house
longhorn" (above). The pupal
chamber is excavated a few centimetres inside the wood and this
causes some damage.

421 Grain-weevil
Sitophilus granarius
L 3–4 mm   All year   Common; in granaries etc.

This serious pest of granaries, mills etc. eats all types of grain. The female bores a hole in the grain using its long rostrum and then lays an egg inside it. The larva feeds on starch and pupates within the hollowed grain. The new adult appears later. Breeding is continuous provided it is not too cold. Where infestations are heavy, the respiratory heat of the weevils may cause the grain to heat and this may destroy it and kill the insects.

422 Dark flour-beetle
Tribolium destructor
L 4–6 mm   All year   Local in buildings

This species is probably of African origin and was introduced into Europe during the thirties. Since then it has spread widely but is not yet of importance in Britain. It can feed on many kinds of organic matter including flour, biscuits, nuts and oilseeds. Like other *Tribolium* species it has glands which produce a foul-smelling liquid. It is sensitive to cold and thrives best in heated buildings. In blocks of flats in can disperse through the ventilation system.

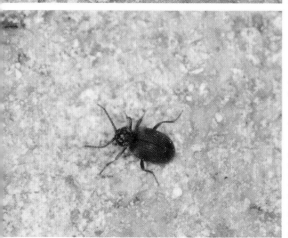

423 Australian spider-beetle
Ptinus tectus
L 2–5 mm   All year   Common; in houses and stores

This species was originally Tasmanian but has now been spread by commerce throughout the temperate world. It occurs in almost every food warehouse and many older private houses in Britain. Its success is due to its having a much higher breeding rate than any of its relatives. Its larvae live on various materials, flour, biscuits, food debris etc. and they can damage packaging and woodwork.

424 Bed-bug
Cimex lectularius
L 4–6 mm　All year　Locally common; in houses etc.

This species is fortunately far less common than it was fifty years ago thanks to higher standards of hygiene and effective control methods. However, the great increase in intercontinental traffic in recent years has caused it to reappear in many places in Europe. The "bed-bug" is a blood sucking hemipteran (true bug). It spends the day hidden behind wallpaper and various fixtures where it leaves dark, round faecal stains. At night it attacks humans and domestic animals. It has a characteristic smell.

425 Silverfish
Lepisma saccharina
L 10–15 mm　All year　Common; indoors

The "silverfish" is a nocturnal insect. It thrives in warm damp places and is therefore found in bathrooms and kitchens. It lives on food remnants expecially crumbs containing sugar. Sometimes it chews at wallpaper to get at the glue underneath, and sometimes it attacks the glue of book bindings. In modern blocks of flats it disperses in the sewage systems where conditions are favourable for it.

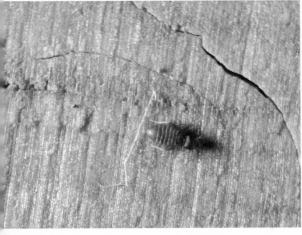

426 Psocid
Liposcelis and other genera
L 1–2 mm　All year　Common; houses and stores

These small pale wingless insects may be seen running about rapidly in old houses and warehouses. They thrive in damp conditions and feed on various scraps of dried animal and vegetable matter including fungi. They can cause damage to books, insect collections and dried plants and sometimes occur in such large numbers in stored cereals that heating occurs (see No 421). In the kitchen pearl barley is often affected. The "book-louse", *Trogium pulsatorium*, can make a noise by banging itself on a surface.

427 Green lace-wing
Chrysopa carnea
L 11–14 mm   All year   Common;
in- and outdoors

This is a crepuscular and nocturnal
insect. In summer it occurs among
bushes and trees where it feeds
largely on honeydew from aphids.
Later in the year it may enter houses
to hibernate and it changes colour
from green to reddish. In spring it
regains its green colour. *Chrysopa*
species can secrete a noxious fluid
from thoracic glands and are
sometimes known as "stink-flies".
Larva No 204.

428 House-cricket (not full-grown)
Acheta domestica
L 14–20 mm   All year   Common; on
refuse dumps, indoors

This species was previously much
commoner in houses than it is today.
In folklore it was identified as a
symbol of household happiness. It
usually occurs in older buildings,
especially where it is warm, such as
bakeries. Unlike grasshoppers crickets are dorso-ventrally flattened with
the wings held horizontally adpres-
sed to the body. The males sing,
mainly at night, their song being produced by the wings being rubbed together. The song consists of a series
of short chirps. Song diagram:

429 House-spider
Tegenaria species
L 9–14 mm   All year   Fairly common; indoors, mainly in cellars and
outhouses.

There are several species of "house-
spider". *T. domestica* is the commonest, while *T. atrica* is the largest.
All of them thrive best indoors
where they spin horizontal webs in
dark nooks and crannies which are
not cleaned too often. The web
catches the occasional fly or other
insect that serve as food for the
spider. They do not normally get
enough peace and quiet in the domestic environment and are therefore more common in sheds and
barns.

430 Rhododendron-whitefly
Dialeurodes chittendeni
L 1–3 mm   June–July   Common;
on rhododendron and azalea

These small relatives of aphids and leaf-hoppers are covered with a white mealy layer of wax. They sit on the undersides of leaves and suck plant sap. They can cause appreciable damage to cultivations when they occur in large numbers. Most species live on wild plants, but some are warmth-loving and occur in greenhouses. One such species is the "greenhouse-whitefly" *Trialeurodes vaporariorum*. The species in the plate occurs mainly on rhododendrons.

431 Palm-thrips
Parthenothrips dracaenae
L 1–2 mm   All year   Common; on potted plants

Thrips are common and troublesome pests of indoor plants. If the leaves of a potted plant turn grey or spotty, look at their undersides. There one often finds adults, larvae and cast skins of thrips. They suck sap from plant cells which become air-filled and dry out. Their feet have bladder-like structures that enable them to walk on smooth surfaces. Many species are common outdoors, e.g. on dandelion flowers. During certain weather conditions thrips become very abundant and often enter houses.

432 Red spider-mite
Tetranychus urticae
L 0.2–0.4 mm   All year   Common; on potted plants, also outdoors

These tiny mites attack a wide range of indoor and outdoor plants. Only the hibernating female is red. They live on the lower surface of leaves where they spin a slight web and suck sap from the cells. Affected leaves turn yellow and fall off. Indoors they thrive in warm dry conditions and spread rapidly from plant to plant. The notorious "fruit tree red spider-mite" is another species, *Metatetranychus ulmi*, but these can also live on fruit trees.

Photographing Insects and other small Animals

To photograph living insects and other small invertebrates successfully one needs good equipment, experience in using it and an adequate knowledge of the habits and environmental requirements of the animals concerned. The best type of camera to use is a single lens reflex as this enables any part of the subject to be brought into sharp focus and it is free from view-finder parallax i.e. one sees through the view-finder exactly what will be recorded on the film. The 6 × 6 cm. format may be used with advantage, especially when the pictures are to be enlarged, but for most purposes miniature (i.e. 35 mm.) cameras are preferable because they are lighter and have greater flexibility in use.

Fairly good close-up photographs of small animals can be taken with relatively few simple additions to the camera when the photographer has plenty of time and patience. Close-up lenses, extension rings, a steady stand, a cable release and an exposure meter should be regarded as essentials. For small subjects bellows can be added and used with or without extension rings and, as these are continuously adjustable, a wide range of magnifications can be obtained. Bellows and rings should, of course, always be equipped for automatic stopping-down of the lens. When extensions are used the amount of light entering the camera is inevitably diminished and exposures have to be lengthened so that even the slightest movements of the subject can cause blurring of the image. This can be overcome by using electronic flash which effectively 'freezes' all movement except very rapid wing-beats. Also, because of the high light intensity produced, the lens can be well stopped-down with consequent improvements in definition and depth of focus.

Modern compact electronic flash-guns have enabled complete hand-held units for this type of work to be bought at modest cost. 'Ring-flash' gives shadow free illumination but is unsuitable for photographing beetles etc. with bright curved surfaces because it produces a circular reflection. For these it is better to use a small rectangular flash-gun preferably in conjunction with a reflecting surface or a second synchronised flash to lighten shadows. The flash-gun should be placed before, and above, or to one side, of the lens and, to avoid having to work very close to the subject, a lens of focal length 100 – 135 mm. rather than the normal one of about 50 mm. should be used. With large and 'shy' insects like butterflies and dragonflies a lens of even greater focal length, say 300 mm. is preferable. Then a very powerful flash will be needed unless the normal one is fitted to a long telescopic extension arm.

For photographing very small animals such as mites and lice where enlargements of X5 to X10 are necessary, it is best to use a microscopic objective lens with a focal length of about 20 mm. This can be mounted on bellows or on a microscope tube and focussing should be carried out by the light of a fluorescent tube which produces much less heat than an incandescent lamp. Stopping-down usually has to be done manually

An effective simple arrangement consists of a single-lens reflex camera with its normal lens, extension rings and a compact electronic flash-gun. At least two extension rings, say 25 mm. & 15 mm. should be available so that the degree of enlargement can be matched to the size of the subject and they should be equipped for automatic stopping-down. The electronic flash should be fitted on a swivelling extension arm so that it can be directed onto the subject and it should not be too powerful because at short range it is easy to over-expose. It is a good idea to take a series of test exposures so as to familiarise oneself with the most suitable stops for different subject distances. A macro-objective with extension ring allowing magnifications of up to 1:1 is more convenient than the normal objective. With this simple set-up one has to work very close to one's subject and the light intensity cannot be varied much.

with this sort of arrangement and the exposure is best made with a pair of synchronised electronic flashes.

Special techniques are necessary when aquatic animals are to be photographed. It is important to avoid reflections at glass and water surfaces and ring-flash should not be used. Shots can be taken from directly above, if the water surface is completely still, or from the side through the glass of a tank. Suitable lighting and screening arrangements can only be found by experimentation.

Even with elaborate equipment ex-perience is needed before satisfactory results can be obtained. One should remember that in this type of work, depth of focus is, at the most, only a few millimetres and, when high magnifica-tions are used, it falls to a fraction of a millimetre. In colour photography, where exposure latitudes are low, it is advisable to take several shots using different apertures, as the correct ex-posure depends, to some extent, on the colour and lightness of the subject. One has to accept that only a proportion of one's shots will prove satisfactory.

An example of a specialised reliable arrangement for insect photography; from left to right: Minicam ring-flash (coupled to accumulator – not shown), Canon lens FLM 100 mm. F4 (a special lens for copying and for insect photography), two Canon extension rings FL 25 mm., Canon bellows FL, one Canon extension ring FL 25 mm., Canon camera FTb. All are mounted on a Linhof universal head which provides a good hand-hold both when carrying and when using the camera. With this arrangement photographs can be taken from a considerable distance (rings removed, bellows collapsed) or from close-up when a magnification of 2:1 can be achieved. A single setting of the aperture can be used (between F11 and F16 with film speed 25 A.S.A and a normally lit subject) because the longer the extension, the shorter the distance between ring-flash and subject – and vice versa.

Systematic Survey

In order to arrange the vast numbers of different animal types into an organised scheme of classification zoologists have put them into **systematic categories.** Animals within a category share characteristics not possessed by those outside it and the categories are ranked at various levels according to the number of characters their members have in common. The system is best explained by a specific example; the Colorado beetle *(Leptinotarsa decemlineata)* is classified thus:

Kingdom — Animalia (all animals)
Phylum — Arthropoda (insects,
             spiders, crabs etc.)
Class   — Insecta (true insects)
Order   — Coleoptera (beetles)
Family — Chrysomelidae (leaf-
             eating beetles)
Genus   — *Leptinotarsa* (a few types
             of latter)
Species — *decemlineata*
             (Colorado beetle)

Intermediate categories can be added e.g. classes may be divided into sub-classes, families may be combined into super-families and sub-species may occur within species.

An ideal classification of animals would be based on evolutionary history and thus reflect the true relationships between the various groups. Practical classifications are, in fact, largely based on supposed evolutionary relationships, but as knowledge is far from complete, subjective criteria influence them to some extent. Certainly in some orders e.g. Lepidoptera and Coleoptera families have been brought together for convenience into groups which are well-known to be artificial. Consequently the systems of classification used in different books vary somewhat in arrangement depending on the opinions of their authors.

The following systematic survey is divided into several sections. The first, pp. 158 – 163 surveys the animal kingdom as a whole. Here animals are not divided beyond sub-class. With a few minor changes (e.g. Hemichordata treated as a phylum) the system follows that of Rothschild (1965). The second section pp. 164 – 165 classifies the insects to orders and occasionally to sub-orders and the third pp. 167 – 175 deals with a selection of the important families within the insect orders Hemiptera, Coleoptera, Hymenoptera, Lepidoptera and Diptera. Here the classification largely follows that given by Imms (1957). The last section, p. 176 deals with a number of families of spiders. Note that whereas the more general surveys, pp. 158 – 165 deal with the world fauna, the subsequent ones relate to N. European forms.

The data on numbers of species within the different groups vary greatly in accuracy and many, particularly those relating to world fauna, have been rounded off, set within likely limits or left open to question. Data from different countries cannot be combined satisfactorily and as yet there is no concensus of opinion as to what constitutes a species and what a sub-species. Even in a well-worked country like Britain the knowledge of different groups varies enormously; e.g. our species of Lepidoptera are known almost completely whereas knowledge of the British Acari is fragmentary. Despite their limitations, however, these data can help the reader to get some idea of the extent of the animal kingdom in Britain and in the world as a whole. The circle diagram on p. 166 gives a rough idea of the comparative numbers of species in various groups in relation to each other and to the entire animal kingdom.

The line drawings of typical animals in the survey summarise the general characteristics of the groups they represent. Note that the scales vary greatly; some species are microscopic whereas others may be several centimetres across.

Notes on systematic tables

The tables are intended to give a broad general view of the animal kingdom. Many groups have, of necessity, been omitted. The following abbreviations are used:

W   estimate of number of species in World

B   estimate of number of species in Britain

no.   followed by number refers to number of plate in colour section

t   following such a number means that reference should be made to text accompanying plate rather than to plate itself

Kingdom *Animalia* Animals

Phylum	W 30,000
Protozoa	B ?
Single-celled animals	

Class	W 6,000
Mastigophora	
Flagellates	no. 388

Class	W 11,500
Rhizopoda	
Amoeboids	

Class	W 3,600
Sporozoa	
Sporozoans	

Class	W 600
Cnidosporidia	

Class	W 6,000
Ciliata	
Ciliates	

Phylum	W 50
Mesozoa	B 3
(Parasitic on marine invertebrates.)	

Phylum	W 5,000
Porifera	B 233
Sponges	

Class	
Calcarea	
Calcareous sponges	

Class	
Hexactinellida	
Glass sponges	

Class	
Demospongiae	
Spongin sponges	
	nos. 376t, 380

Phylum	W 9,600
Cnidaria	B 260+
Coelenterates (Animals with stinging cells.)	

Class	W 2,700
Hydrozoa	B 150+
Hydroids & hydromedusae	
	no. 382

Class	W 200
Scyphozoa	B 50+
Jellyfish	nos. 383–384

Class	W 6,000–6,500
Anthozoa	B 59
Sea anemones, corals	
	no. 381

Phylum	W 80–100
Ctenophora	B 3
Sea gooseberries	

Phylum W 15,000
Platyhelminthes
Flatworms

Class W 1,600+
Turbellaria B 100
Planarians no. 358

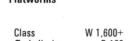

Class W 2,400+
Trematoda B 500+?
Flukes (parasitic) no. 327t

Class W 1,500+
Cestoda B 500+?
Tapeworms (parasitic)

Phylum W 750
Nemertina B 55
Ribbon worms
(Free-living, mainly marine)

Phylum W 12,000+
Aschelminthes
Unsegmented worms

Class W 1,500
Rotifera B 200–300
Wheel animalcules
(aquatic)

Class W 10,000
Nematoda B 1,000+?
Roundworms no. 66
(Free-living or parasitic)

Class W 200
Gastrotricha B 12
(Tiny aquatic ciliated
animals)

Class W 250
Nematomorpha B 4
Horse-hair worms (adults
aquatic, larvae parasitize
insects)

Class W 100
Echinoderida B 11
(Small marine crawling
forms)

Class W 8
Priapulida B 2
(Mud-dwelling marine
"worms")

Phylum W 600?
Acanthocephala B 30?
Thorny-headed worms
(parasitic)

Phylum W 60
Entoprocta B 9?
(Stalked hydroid like aquatic
forms)

Phylum W 8,000
Annelida B 600
Segmented worms

Class W 5,000
Polychaeta B 260+
Bristle worms
nos. 376t, 385–387

Class W 2,500
Oligochaeta B 170+
Earthworms etc.
nos. 284, 357

Class W 300
Hirudinea B 28
Leeches nos. 359–360

Phylum W 250
Sipuncula B 11
Peanut worms
(marine bottom-dwellers)

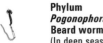

Phylum W 70–150
Echiura B 6
(Marine, often burrowing
forms)

Phylum W 50–100
Pogonophora B 7
Beard worms
(In deep seas)

Phylum
Arthropoda

Sub-class   W 18,000
Malacostraca

Class
Crustacea

Order   W 7
Nebaliacea   B 1

Sub-class
Branchiopoda   B 100
(Mostly fresh-water forms)

Order   W 350
Mysidacea   B 74
Opossum-shrimps

Order   W 4
Cephalocarida   B 0?

Order   W 400
Cumacea   B 72

Order   W 1,000
Cladocera   B 90
Water fleas   no. 354

Order   W 300
Tanaidacea   B 10?

Order   W 200
Anostraca   B 2
Fairy shrimps

Order   W 4,000
Isopoda   B 120+
Woodlice etc.
   nos. 279, 280t, 355, 395

Sub-class   W 20,000
Ostracoda   B 300?
Ostracods   no. 352

Order   W 3,600
Amphipoda   B 200+
Sand-hoppers etc.
   nos. 356, 394, 396

Sub-class   W 4,500
Copepoda   B 880
Copepods   no. 353

Order   W 90
Euphausiacea   B 3?
Krill, whale-feed

Sub-class   W 3
Mystacocorida   B 0?

Sub-class   W 75
Branchiura   B 1?
Carp-lice

Order   W 8,000
Decapoda   B 91
Crabs, lobsters etc.
   nos. 391–393

Sub-class   W 800
Cirripedia   B 29
Barnacles   no. 390

Class | W 8,000
Diplopoda | B 44
Millipedes | nos. 280–281

Class | W 60
Pauropoda | B 47
(Small soil forms, branched antennae)

Class | W 2,000–3,000
Chilopoda | B 44
Centipedes | nos. 282–283

Class | W 120
Symphyla | B 14
White centipedes
(plant eaters)

See pages 164–176

Class | W 760–810,000
Insecta | B 20,000+
Insects

Class | W 5
Merostomata | B 0
King-crabs

Class | W 80–90,000
Arachnida | B 2,450?

Order | W 1,100–1,500
Pseudoscorpiones | B 26
False scorpions

See page 176

Order | W 40–50,000
Araneae | B 584
Spiders

Order | W 3,000
Opiliones | B 22
Harvestmen | no 277

Order | W 30,000
Acari | B 1,800?
Mites & Ticks | nos. 72,
210, 278, 339, 408, 432

Class | W 500
Pycnogonida | B 21
Sea spiders

Class | Extinct
Trilobita
Trilobites (In rock 225 to
570 million years old)

Class | W 120
Onychophora | B 0
Velvet worms (non-
European primitive
arthropods)

Class | W 60
Pentastomida | B 1?
(Small elongate clawed
parasites)

Class | W 180
Tardigrada | B 50?
Water-bears
(In moss, fresh & salt-water)

Phylum W 100,000
Mollusca B 800

Class W 2
Monoplacophora B 0
(Very rare Pacific deep-sea forms)

Class W 1,000
Polyplacophora B 12
Chitons

Class W 150
Aplacophora B 5
(Worm-like, shell-less marine forms)

Class W 100,000?
Gastropoda B 520+
Snails etc.

Sub-class W 60,000?
Prosobranchia B 245
Sea snails etc.
nos. 364–369

Sub-class W 10,000?
Opisthobranchia B 140+
Sea slugs etc.

Sub-class W 30,000
Pulmonata B 136
Land & freshwater snails
nos. 148–150, 207, 285,
327, 361–362

Class W 350
Scaphopoda B 5
Tusk shells (marine)

Class W 20–25,000
Bivalvia B 200+
Bivalves nos. 363, 370–377

Class W 700
Cephalopoda B 19
Octopods, squids etc.

Phylum W 18
Phoronida B 4
(Worm-like tube dwellers with tentacles)

Phylum W 4,000
Polyzoa B 290
Moss animals no. 389

Phylum W 260
Brachiopoda B 15–20
Lamp shells
(bivalves, often stalked)

Phylum W 50
Chaetognatha B 18
Arrow worms
(marine planktonic)

Phylum W 91
Hemichordata B 6

Class W 20
Pterobranchia B 1
(Small stalked, often colonial marine forms)

Class W 70
Enteropneusta B 3
Acorn worms
(shore burrowing forms)

Phylum W 5,700
Echinodermata B 170+
(Spiny skinned animals)

Class W 80
Crinoidea B 12
Sea lilies, feather stars

Class W 1,600
Asteroidea B 47
Starfish no. 378

Class W 1,500–2,000
Ophiuroidea B 48
Brittle stars, basket stars

Class W 800
Echinoidea B 34
Sea urchins no. 379

Class W 600–1,000
Holothuroidea B 30
Sea cucumbers

Phylum
Chordata
Chordate animals

Sub-phylum W 1,600
Tunicata B 75?
Tunicates (marine forms)

Class W 1,200
Ascidiacea B 61
Sea squirts (sedentary)

Class W 30
Larvacea B 3
(Neotenic pelagic forms)

Class W 30
Thaliacea B 5
Salps etc.
(pelagic forms)

Sub-phylum W 20
Acrania B 1
Lancelets
(coastal sand-dwellers)

Sub-phylum
Vertebrata
Animals with backbones

Class W 50
Agnatha B 3
Lampreys & hagfishes

Class W 25–30,000
Pisces B 350
Bony fishes

Class W 3,000–3,500
Amphibia B 8
Amphibians

Class W 5,000–6,000
Reptilia B 10
Reptiles

Class W 8,700
Aves B 467
Birds

Class W 4,200
Mammalia B 95
Mammals

In addition to forms alive at present a great many species are known only as fossils. Most of these belong to groups whose members have hard shells or bony skeletons. In the following table estimates of the numbers of known fossil species in selected groups are given.

	Number of known fossil species
Rhizopoda — amoeboid animals, mostly foraminifera	15,000
Trilobita — trilobites	10,000
Mollusca	150,000
' *Polyplacophora* — chitons	100
" *Gastropoda* — snails etc.	15,000
" *Cephalopoda* — octopods etc.	10,000
Polyzoa — moss animals	15,000
Brachiopoda — lamp shells	30,000
Echinodermata	45,000

Some authorities believe that the true number of fossil species may lie between 300 and 1,000 millions.

Class *Insecta* Insects

Sub-class *Apterygota*

Order	W 1,000–2,000
Collembola	B 300+
Spring-tails	no. 351

Order	W 140?
Protura	B 12

(Minute antenna-less soil forms)

Order	W 400–500
Diplura	B 12

Two-prong bristle-tails

Order	W 400
Thysanura	B 9
Bristle-tails	no. 425

Sub-class *Pterygota (Exopterygota)*

Order	W 1,400
Ephemeroptera	B 46

Mayflies
nos. 310–311, 343–344

Order	W 4,500
Odonata	B 43

Dragonflies

Sub-order	
Zygoptera	B 17

Damselflies
nos. 301–309, 342

Sub-order	
Anisoptera	B 25

True dragonflies
nos. 286–300, 341

Order
Dictyoptera

Sub-order	W 3,500
Blattodea	B 3
Cockroaches	no. 123

Sub-order	·W 1,800
Mantodea	B 0
Mantises	

Order	W 1,900
Isoptera	B 0
Termites	

Order	W 20
Zoraptera	B 0

(Obscure small foreign forms)

Order	W 1,600
Plecoptera	B 35
Stoneflies	nos. 313, 345

Order	W 150
Embioptera	B 0

Web-spinners
(In warm countries only)

Order	W 6
Grylloblattodea	B 0

Rock crawlers
(wingless foreign forms)

Order	W 20,000
Orthoptera	B 30

Grasshoppers & crickets

Super-family	W 5,000
Tettigonioidea	B 10

Bush-crickets etc.
nos. 124–126, 205–206

Super-family	W1,200
Grylloidea	B 4

Crickets etc.
nos. 274, 428

Super-family	W 8,000
Acridoidea	B 14

Grasshoppers etc.
nos. 127–129, 319–320

Order	W 2,000
Phasmida	B 0

Stick & leaf insects

Order	W 900
Dermaptera	B 4
Earwigs	no. 241

Order	W 1,000
Psocoptera	B 70

Psocids & book-lice
no. 426

	Order	W 250
	Siphunculata	B 25+
	Sucking lice	no. 409

	Order	W 5,000–6,000
	Mallophaga	B 500+
	Biting lice	no. 410

	Order	W 2,000–3,000
	Thysanoptera	B 160
	Thrips	nos. 431

	Order	W 55,000
See page 167	*Hemiptera*	B 1,650+
	True bugs	

Sub-class *Pterygota (Endopterygota)*

	Order	W 300,000
See pages 168–169	*Coleoptera*	B 4,000+
	Beetles	

	Order	W 300
	Strepsiptera	B 18
	Stylops	

	Order	W 150,000
See pages 170–171	*Hymenoptera*	B 6,200+
	Ants, bees, wasps etc.	

	Order	W 4,000
	Neuroptera	B 60
	Lace-wings, ant-lions etc.	
	nos. 204, 267, 312, 427	

	Order	W 350
	Mecoptera	B 4
	Scorpion flies etc.	
	nos. 187, 238	

	Order	W 4,000
	Trichoptera	B 190+
	Caddisflies	
	nos. 314, 346	

	Order	W 120,000–150,000
See pages 172–173	*Lepidoptera*	B 2,469
	Butterflies & moths	

	Order	W 80,000–90,000
See pages 174–175	*Diptera*	B 5,200
	Two-winged flies	

	Order	W 1,800
	Siphonaptera	B 60
	Fleas	no. 411

Diagram to show the relative numbers of species in different animal groups.

About 1,200,000 species of animals are actually known to live in the world at the present time. The following diagram shows how this number is distributed between different animal groups. Arthropods are clearly most numerous and amongst them insects dominate.

The true number of species will greatly exceed the known number, because of the multitudes of small species and of species from inaccessible localities which have not yet been discovered. Possibly the true total will be somewhere between 5 and 10 million species. Something like 10,000 new species and sub-species are described every year.

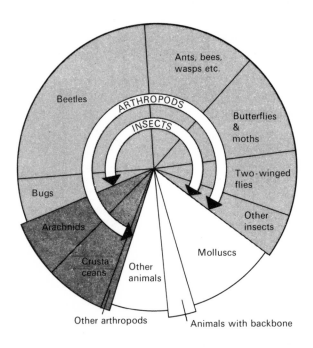

Order *Hemiptera* True bugs

Sub-order *Homoptera* Plant bugs

Super-family *Cicadoidea* Leaf-hoppers etc. nos. 117, 203, 321	W 24,000 B 270
Super-family *Coccoidea* Scale-insects & mealy-bugs no. 75	W 5,000 B 141
Super-family *Aphidoidea* Greenfly etc. nos. 62–63, 67, 74, 76, 236	W 3,800 B 526
Super-family *Psylloidea* Jumping plant-lice no. 77	W 1,100–1,200 B 80
Super-family *Aleyrodoidea* Whiteflies no. 430	W 200? B 15

Family *Gerridae* Pondskaters no. 332	W 200 B 10
Family *Veliidae* Water-crickets no. 332t	W 100 B 5
Family *Nabidae* Damsel bugs no. 118	W 350 B 12
Family *Tingidae* Lacebugs	W 700 B 23
Family *Aradidae* Flatbugs	W 500 B 5
Family *Anthocoridae* Flower bugs	W 450 B 28
Family *Lygaeidae* Groundbugs no. 122	W 2,000 B 74
Family *Coreidae* Squash bugs	W 2,000 B 10
Family *Cimicidae* Bed-bugs etc. no. 424	W 70 B 3
Family *Pyrrhocoridae* Firebugs no. 122t	W 450 B 1
Family *Miridae* Capsid bugs nos. 31, 119	W 5,000 B 203
Family *Pentatomidae* Shieldbugs nos. 120–121, 202	W 5,000? B 20

Sub-order *Heteroptera*

Family *Corixidae* Water-boatmen no. 335	W 300 B 33
Family *Notonectidae* Backswimmers no. 334	W 200 B 4
Family *Nepidae* Waterscorpion no. 337	W 200 B 2
Family *Naucoridae* Saucer-bugs no. 336	W 200 B 1
Family *Hydrometridae* Water-measurers	W 70 B 2

Order *Coleoptera* Beetles

Sub-order *Adephaga*
Super-family *Caraboidea*

Family	W 2,000
Cicindelidae	B 5
Tiger beetles	nos. 248, 249

Family	W 20–25,000
Carabidae	B 350
Ground beetles	
	nos. 162t, 243–247, 399

Family	W 200
Haliplidae	B 18
(Small aquatic alga-feeding	
species)	

Family	W 4,000
Dytiscidae	B 107
True water beetles	
	nos. 329–330, 340

Family	W 400–500
Gyrinidae	B 11
Whirligig beetles	no. 331

Sub-order *Polyphaga*
Super-family *Hydrophiloidea*

Family	W 2,000
Hydrophilidae	B 120
	no. 328
(Long-palped aquatic and	
land forms)	

Super-family *Histeroidea*

Family	W 2,000
Histeridae	B 39
(Scavengers in dung,	
carrion etc.)	

Super-family *Staphylinoidea*

Family	W 2,000
Silphidae	B 60
Carrion beetles	
	nos. 243–255

Family	W 3,500
Pselaphidae	B 30
Ant beetles	no. 265t

Family	W 20,000
Staphylinidae	B 1,000
Rove beetles	nos. 242, 318

Super-family *Scarabaeoidea*

Family	W 20,000
Scarabaeidae	B 77
Chafers, stag & dung-	
beetles	nos. 35–36, 109,
	190–194, 250–252, 272

Super-family *Byrrhoidea*

Family	W 270
Byrrhidae	B 10
Pill beetles	

Super-family *Buprestoidea*

Family	W 10–15,000
Buprestidae	B 12
(Shining metallic adults,	
usually wood-boring	
larvae)	

Super-family *Elateroidea*

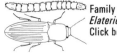

Family	W 8,000
Elateridae	B 65
Click beetles & wire worms	
	nos. 110–111, 243

Super-family *Cantharoidea*

Family	W 2,000
Lampyridae	B 2
Glow worms, fire flies	
	no. 112

Family	W 6,000
Cantharidae	B 41
Soldier beetles etc.	
	nos. 32, 240

Super-family *Dermestoidea*

Family	W 700
Dermestidae	B 25
Museum beetle, bacon	
beetle etc.	nos. 417–418

Super-family *Bostrychoidea*

Family	W 1,200
Anobiidae	B 31
Wood worms, death watch	
etc.	

Family	W 500
Ptinidae	B 20
Spider beetles	no. 423

Super-family *Cleroidea*

| Family *Cleridae* | W 3,000 B 12 no. 225 |
| Family *Melyridae* | W 4,000 B 23 |

Super-family *Lymexyloidea*

| Family *Lymexylidae* | W 50 B 2 |

Super-family *Cucujoidea*
Section *Clavicornia*

Family *Nitidulidae*	W 2,500 B 92 no. 58
Family *Rhizophagidae*	B 11
Family *Cucujidae*	B 32
Family *Cryptophagidae*	W 800 B 80
Family *Byturidae* (Raspberry beetle etc.)	B 3 no. 69
Family *Coccinellidae* Ladybirds	W 5,000? B 45 nos. 106–108
Family *Lathridiidae*	W 600 B 50

Section *Heteromera*

| Family *Tenebrionidae* | W 17,000 B 35 no. 422 |
| Family *Lagriidae* | W 600 B 1 |

Family *Pythidae*	B 1
Family *Pyrochroidae* Cardinal beetles	W 150 B 3 no. 195
Family *Mordellidae*	W 1,100 B 24
Family *Rhipiphoridae*	W 400+ B 1
Family *Meloidae* Oil beetles, blister beetles	W 2,300 B 9 no. 113
Family *Oedemeridae*	W 600 B 7

Super-family *Chrysomeloidea*

Family *Cerambycidae* Longhorn beetles	W 20–30,000 B 70 nos. 32–33, 37, 196–197, 220–224, 419–420
Family *Bruchidae*	W 1,000 B 15
Family *Chrysomelidae* Leaf beetles	W 30,000 B 250+ nos. 60, 64, 115–116, 198–200, 316–317

Super-family *Curculionoidea*

Family *Anthribidae*	W 2,000 B 10
Family *Curculionidae* Weevils	W 50,000 B 500+ nos. 59, 78, 114, 201, 211–212, 226–227, 421
Family *Scolytidae* Bark beetles	W 2,000 B 57 nos. 228–229

Order *Hymenoptera*

Sub-order *Symphyta*

Super-family *Megaladontoidea* — W 200

Family — W 160
Pamphiliidae — B 20
(Sawflies found on trees)

Super-family *Siricoidea* — W 140

Family — W 70
Siricidae — B 4
Wood wasps — no. 231

Super-family *Cephoidea* — W 100

Family — W 100
Cephidae — B 12
Stem sawflies
(cereals & shrubs)

Super-family *Tenthredinoidea*

Family — W 500
Argidae — B 14
(Sawflies on shrubs) — no. 73

Family — W 130
Cimbicidae — B 12
(Sawflies on shrubs e.g.
hawthorn)

Family — W 60
Diprionidae — B 9
(Sawflies on conifers)
no. 230

Family — W 4,500
Tenthredinidae — B 400
(Sawflies on many plants)
nos. 71, 79, 132

Sub-order *Apocrita*
Division *Parasitica* — W 101,000

Super-family *Ichneumonoidea*

Family — W 60,000
Ichneumonidae — B 1900+
Ichneumon flies (parasitic
esp. on insects)
nos. 130–131, 231t, 233

Family — W 10,000
Braconidae — B 790+
(Parasitic on insects)

Super-family *Cynipoidea* — W 2,500

Family — W 1,400
Cynipidae — B 200
Gall wasps
nos. 208–209, 213

Family — W 200
Eucoilidae — B 58+
(Parasitic wasps)

Super-family *Chalcidoidea* Chalcid wasps

Family — W 1,100
Mymaridae — B 32
Fairy flies
(Parasites on insect eggs)

Family — W 360
Trichogrammatidae — B 13
(Small parasites on insect
eggs)

Family — W 600
Aphelinidae — B 38
(Parasites esp. on
Homoptera)

Family — W 1,000
Chalcididae — B 9
(Parasites esp. on
Lepidoptera & Diptera)

Family — W 1,000
Torymidae — B 87
(Parasites esp. on gall-
insects)

Family — W 1,000
Eurytomidae — B 77
(Mostly attacking plants)

Family — W 3,500
Eulophidae — B 480+
(Parasites esp. on leaf-
miners)

Family — W 2,000
Encyrtidae — B 144
(Parasitic on insects)

Family — W 4,000
Pteromalidae — B 490+
(Parasitic on insects)

Super-family *Proctotrupoidea*   W 7,000

    Family   W 1,500
    Platygasteridae   B 146
    (Parasites esp. on gall-
    midges)

    Family   W 1,800
    Diapriidae   B 120+
    (Parasites esp. on Diptera)

    Family   W 100
    Proctotrupidae   B 33
    (Parasites esp. on
    Coleoptera)

    Family   W 600
    Ceraphrontidae   B 100+
    (Mostly secondary parasites
    on parasites of Homoptera)

Division *Aculeata*   W 38,800

  Super-family *Bethyloidea*   W 1,600

    Family   W 1,500
    Chrysididae   B 21
    Cuckoo or Ruby-tailed wasps
    no. 261

  Super-family *Scolioidea*   W 4,500

    Family   W 1,000
    Scoliidae   B 0
    (Large ectoparasites on
    Coleoptera)

    Family   W 3,000
    Mutillidae   B 2
    Velvet "ants"
    (not really ants!)

  Super-family *Formicoidea*   W 12,500

    Family   W 12,500
    Formicidae   B 36
    Ants   nos. 262–266

  Super-family *Pompiloidea*   W 3,100

    Family   W 3,100
    Pompilidae   B 40
    Spider-hunting wasps
    no. 256

Super-family *Vespoidea*   W 1,000

    Family   W 300
    Vespidae   B 27
    Wasps & hornets
    nos. 38–39

Super-family *Sphecoidea*   W 5,100

    Family   W 5,000
    Sphecidae   B 230
    Digger and Sand wasps
    nos. 43, 257–259

Super-family *Apoidea* Bees   W 11,000

    Family   W 1,500
    Apidae   B 60+
    (Social & solitary bees)
    nos. 40–42

    Family   W 1,000
    Andrenidae   B 60+
    (Mining bees etc.)   no. 260

    Family   W 2,0000
    Halictidae   B 50+
    (Mining bees & their
    parasites)

    Family   W 2,000
    Megachilidae   B 30+
    (Leaf-cutter, mason bees
    etc.)

Order *Lepidoptera* Moths and butterflies

Sub-order *Zeugloptera*

Family	W 100
Micropterigidae	B 5

(Moths with mandibles but no proboscis)

Sub-order *Monotrysia*

Super-family *Hepialoidea*

Family	W 300
Hepialidae	B 5
Swift moths	no. 89

Super-family *Nepticuloidea*

Family	W 300
Nepticulidae	B 77

(The smallest moths)

Super-family *Incurvarioidea*

Family	
Incurvariidae	B 25

Bright & longhorn moths

no. 186

Sub-order *Ditrysia*

Super-family *Cossoidea*

Family	W 500?
Cossidae	B 3

Goat & leopard moths

nos. 171, 271

Super-family *Zygaenoidea*

Family	W 900
Zygaenidae	B 10

Burnets & foresters

nos. 25–26

Family	W 1,000
Limacodidae	B 2

(Moths with slug-like larvae)

Super-family *Tineoidea*

Family	W 400
Psychidae	B 23?
Bagworms	no. 104

Family	W 2,400
Tineidae	B 56

Clothes & corn moths etc.

Super-family *Yponomeutoidea*

Family	W 800
Sesiidae	B 15
Clearwings	

Family	W 200
Yponomeutidae	B 50

Small ermine moths

nos. 55, 81, 185

Super-family *Gelechioidea*

Family	W 900
Coleophoridae	B 80

Casebearer moths

Family	W 200+
Elachistidae	B 40

(Larvae mine grasses etc.)

Family	W 4,000
Gelechiidae	B 150

Nebs or groundlings

Super-family *Tortricoidea*

Family	W 4,500
Tortricidae	B 350

Tortrix moths

nos. 80, 183, 235

Super-family *Pyraloidea*

Family	W 10,000
Pyralidae	B 160

Meal moths etc.

nos. 97–98, 219, 315

Super-family *Pterophoroidea*

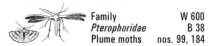

Family	W 600
Pterophoridae	B 38
Plume moths	nos. 99, 184

Super-family *Hesperioidea*

Family	W 3,000
Hesperiidae	B 8
Skippers	nos. 19–21

Super-family *Papilionoidea*

Family	W 600
Papilionidae	B 1
Swallow tails	nos. 1, 2, 100

Family   W 1,000–1,500
Pieridae   B 10
Whites etc.
        nos. 3–6, 52–54

Family   W 3,000
Lycaenidae   B 17
Blues, coppers, hairstreaks
etc.
       nos. 17–18, 22–24, 154

Family   W 1,000
Nemeobiidae   B 1

Family   W 2,000?
Nymphalidae   B 19
(Many familiar species)
       nos. 7–16, 101

Family   W 2,000
Satyridae   B 11
Browns etc.   nos. 82–87

Super-family *Bombycoidea*

Family   W 1,200
Lasiocampidae   B 11
Eggars, lappet moths no. 103

Family   W 1,200
Saturniidae   B 1
Emperor moth etc.
       nos. 166–167

Family   W 1
Endromidae   B 1
(One large day-flying
species)

Super-family *Geometroidea*

Family   W 350?
Drepanidae   B 6
Hook tips   no. 170

Family
Thyatiridae   B 9
(Robust moths resembling
noctuids)

Family   W 12-15,000
Geometridae   B 250+
Geometer moths & looper
caterpillars
     nos. 70, 91–96, 175–182,
                     218

Super-family *Sphingoidea*

Family   W 1,000
Sphingidae   B 18
Hawk moths   nos. 27–28,
    102, 155–158, 214–215

Super-family *Notodontoidea*

Family   W 2,000
Notodontidae   B 24
Prominents etc.
       nos. 160–161, 163–164

Family   W 80
Thaumetopoeidae   B 0
(Moths with "process-
ionary" caterpillars)

Super-family *Noctuoidea*

Family   W 20–30,000
Noctuidae   B 290+
Noctuid moths
  nos. 29–30, 65, 90–91, 105,
       168, 172–174, 217

Family   W 2,000
Lymantriidae   B 10
Tussock moths
       nos. 159, 162, 216

Family   W 5,000–6,000
Arctiidae   B 32
Tiger moths etc.
       nos. 88, 165, 169

Family   W 2,000
Ctenuchidae   B 0
(Sawfly- or burnet-like
moths)

Family
Nolidae   B 5

Order *Diptera* Two-winged flies

Sub-order *Nematocera*

Family
Bibionidae
March flies
W 400
B 18
no. 136

Family
Mycetophilidae
Fungus gnats
W 1,100
B 400

Family
Sciaridae
Mushroom flies etc.
W 500
B 80
no. 416

Family
Cecidomyiidae
Gall midges
nos. 57, 153, 237
W 5,000
B 600+

Family
Trichoceridae
Winter gnats
W 100
B 10
no. 239

Family
Psychodidae
Moth flies or owl midges
no. 324
W 400
B 70

Family
Culicidae
Mosquitos & gnats
nos. 350, 405
W 2,400
B 30

Family
Chaoboridae
(Midge like flies)
B 6
no. 348

Family
Simuliidae
Black flies
W 1,000
B 19
nos. 349, 406

Family
Ceratopogonidae
Biting midges
W 600
B 130
no. 407

Family
Chironomidae
Non-biting midges
nos. 323, 347
W 3,000
B 400

Family
Tipulidae
Crane flies or daddy-long-legs
nos. 133–135, 322
W 12,000

Sub-order *Brachycera*

Family
Stratiomyidae
Soldier flies
W 1,500
B 50+

Family
Rhagionidae
Snipe flies
W 500
B 21
no. 189

Family
Tabanidae
Horse flies, clegs etc.
nos. 400–402
W 3,000
B 30

Family
Therevidae
Stiletto flies
W 500
B 10

Family
Scenopinidae
Window flies
W 50
B 3

Family
Asilidae
Robber flies
nos. 139–141, 232
W 5,000
B 26

Family
Cyrtidae
(Parasitic as larvae on spiders)
W 200
B 3

Family
Bombyliidae
Bee flies
W 3,000
B 12
no. 45

Family
Empididae
(Predacious flies which form "dancing" aerial swarms)
no. 188
W 3,000
B 300+

Family
Dolichopodidae
Long-legged flies
W 3,000
B 250+
no. 325

Sub-order *Cyclorrhapha*

Family
Platypezidae
Flat-footed flies
W 100
B 23

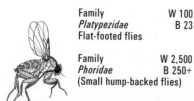

Family
Phoridae
(Small hump-backed flies)
W 2,500
B 250+

Family   W 4,000
Syrphidae   B 250
Hover flies   nos. 46–50

Family   W 400
Pipunculidae   B 40
(Small hovering flies, larvae
parasitic on Homoptera)

Family   W 20
Lonchopteridae   B 7
(Pointed-winged flies;
larvae in leaf litter)

Family   W 500
Conopidae   B 18
(Bee or wasp-like flies;
larvae parasitic esp. on
adult Hymenoptera)

Family   W 100
Psilidae   B 28
Carrot fly & allies no. 68

Family   W 2,000
Trypetidae   B 70
Fruit flies

Family   W 1,000
Agromyzidae   B 90
Small flies; larvae mine
leaves   no. 152

Family   W 250
Helomyzidae   B 63
(Flies with scavenging
larvae)   no. 138

Family   W 70
Piophilidae   B 9
Cheese skipper etc.

Family   W 200
Sciomyzidae   B 50+
(Larvae parasitic on snails
etc.)

Family   W 150
Sepsidae   B 23
(Ant-like flies;
larvae saprophagous)
             no. 137

Family   W 20
Coelopidae   B 7
Kelp flies   no. 397

Family   W 1,000
Drosophilidae   B 31
Vinegar flies etc. no. 415

Family   W 200
Sphaeroceridae B 100
(Small flies; saprophagous
larvae)

Family   W 1,000
Ephydridae   B 120
Shore flies (sea & lake)

Family   W 1,000
Chloropidae   B 80+
Frit fly, gout fly etc.

Family   W 3
Braulidae   B 1
Bee louse etc.

Family   W 100
Hippoboscidae B 9
Sheep ked, forest fly etc.

Family   W 500
Cordiluridae   B 50+
Dung flies etc. no. 270

Family   W 5,000
Muscidae   B 450+
House flies, cabbage root
flies etc.   nos. 56, 61, 151,
    269, 398, 403–404, 412

Family   W 2,000
Calliphoridae B 100
Blow flies
     nos. 51, 268, 413–414

Family   W 4,000
Tachinidae   B 250+
(Larvae parasitic on insects)

Family   W 100
Oestridae   B 6
Warble flies, sheep nostril
fly

Order *Araneae* Spiders

Family
Atypidae   B 1
Purse web spiders

Family
Salticidae   B 33
Jumping spiders   no. 276

Family
Thomisidae   B 37
Crab spiders   no. 44

Family
Sparassidae   B1
Crab spiders   no. 146

Family
Clubionidae   B 34

Family
Zoridae   B 4

Family
Gnaphosidae   B 27

Family
Oxyopidae   B 1
Lynx spiders

Family
Lycosidae   B 37
Wolf spiders   no. 275

Family
Pisauridae   B 3
Wolf spiders   nos. 147, 326

Family
Argyronetidae   B 1
Water spider   no. 338

Family
Agelenidae   B 22
no. 429

Family
Mimetidae   B 3

Family
Tetragnathidae   B 9
no. 144

Family
Argiopidae   B 38
nos. 142–143

Family
Linyphiidae   B 250
no. 234

Family
Dysderidae   B 6
Six eyed spiders

Family
Nesticidae   B 2
Comb footed spiders

Family
Theridiidae   B 48
Comb footed spiders   no. 145

Family
Dictynidae   B 16
Mesh webbed spiders

Family
Uloboridae   B 2
Lace web spiders

Anatomy of Insects and Spiders

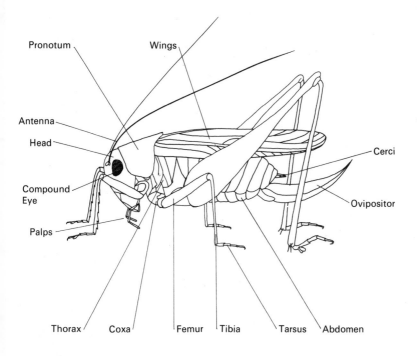

Pronotum
Wings
Antenna
Head
Compound Eye
Palps
Cerci
Ovipositor
Thorax Coxa Femur Tibia Tarsus Abdomen

Insect: the body is divided into three sections and there are three pairs of legs.

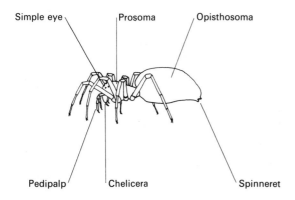

Simple eye
Prosoma
Opisthosoma
Pedipalp
Chelicera
Spinneret

Spider: the body is divided into two parts and there are four pairs of legs.

Insect development, metamorphosis and classification

Apterygota
Primitively wingless
insects. Metamorphosis
slight.
e.g. silverfish no. 425

Pterygota
Winged (and secondarily wingless) insects

Exopterygota
Metamorphosis slight,
wings develop externally
e.g. grasshopper no. 320

Endopterygota
Metamorphosis complex,
wings develop internally
e.g. butterfly no. 1

Egg

Egg

Egg

Larval stages (nymphs)

Larval stages (nymphs)

Larval stages

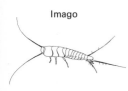

Pupa

Imago

Imago

Imago

Insect orders within the above groups:

Thysanura – Bristle-tails	Ephemeroptera – Mayflies	Neuroptera – Lace wings etc.
Diplura – 2-prong bristle-tails	Odonata – Dragonflies etc.	Mecoptera – Scorpion flies
Protura	Plecoptera – Stoneflies	Lepidoptera – Butterflies etc.
Collembola – Spring-tails	Grylloblattodea	Trichoptera – Caddis flies
	Orthoptera – Grasshoppers etc.	Diptera – 2-winged Flies
	Phasmida – Stick insects etc.	Siphonaptera – Fleas
	Dermaptera – Earwigs	Hymenoptera – Ants, bees etc.
	Embioptera – Web-spinners	Coleoptera – Beetles
	Dictyoptera – Cockroaches etc.	Strepsiptera – Stylops
	Isoptera – Termites	
	Zoraptera	
	Psocoptera – Booklice etc.	
	Siphunculata – Sucking lice	
	Mallophaga – Biting lice	
	Hemiptera – True bugs	
	Thysanoptera – Thrips	

Bibliography

The bibliography has been divided into the following three sections:

I. general works dealing with the environments considered in the colour-plate section of the book.

II. works on invertebrates in general and general accounts of separate groups.

III. works dealing with the collection, preservation and identification of various animal groups.

Section I

On flowers and fruit.

Ministry of Agriculture, Fisheries and Food advisory leaflets deal with many species found in these environments.

Becker, P. 1974 *Pests of Ornamental Plants,* Bulletin no. 97, (Ed. 3) Ministry of Agriculture, Fisheries and Food. H.M.S.O. London. Pests of flowers, shrubs, etc. Line drawings, monochrome photographs, short bibliography.

Massee, A.M. 1954. *The Pests of Fruits and Hops* (Ed. 3) Crosby Lockwood, London. Life-histories and control measures for many pest species. Beneficial and harmless species, insecticides and spraying machines. Monochrome photographs, references.

On cultivated plants.
On grass and herbs.

Ministry of Agriculture, Fisheries and Food advisory leaflets deal with many species here.

Anon. 1965. *Cereal Pests.* Ministry of Agriculture, Fisheries and Food, Bulletin 186. Pests of growing cereal crops described with details of damage and control. Black and white and colour illustrations.

Edwards, C.A. & Heath, G.W., 1964. *The Principles of Agricultural Entomology.* Chapman & Hall, London. A general handbook on arthropod farm pests. Descriptions and life-cycles of many species. Keys to species of pests on farm animals, on farm-stored grain, on cereals, on fodder-crops, root crops and vegetables. Information on pest control. Line drawings, monochrome photographs, many references.

Gram, E., Bovien, P. and Stapel, C. 1969. *Recognition of Diseases and Pests of Farm Crops* Blandford, London. Colour pictures and captions for several hundred disease and pest conditions.

Jones, F.G.W. and Dunning, R.A. 1972. *Sugar Beet Pests* Bulletin 162 (Ed. 3). Ministry of Agriculture, Fisheries and Food. H.M.S.O. London. Insects, nematodes, birds and mammals affecting sugar beet.

Jones, F.G.W. and Jones, M.G. 1974. *Pests of Field Crops* (Ed. 2), Arnold, London. Text-book on animal pests of farm crops. Includes arthropods, molluscs, nematodes, birds and mammals. Much detailed biology. Sections on pests of storage, pest management, pest control chemicals. Line drawings, monochrome photos, colour plates. Many references.

On deciduous trees and bushes.
On conifers.

Ministry of Agriculture, Fisheries and Food advisory leaflets deal with many species here.
Pests of ornamental plants and of fruit may occur here too.

Chrystal, R.N. 1937. *Insects of the British Woodlands* Warne, London. General account of woodland and forestry species. Systematic coverage also classification by habits. Keys to some groups. Line drawings. Monochrome photos. Selective bibliography.

On and under the ground.

Kevan, D.K. McE., 1962. *Soil Animals* Witherby, London. Systematics and ecology of soil animals. Sampling and extraction methods, adaptations of animals to soil life and effects of animal activity on soil are discussed. Text diagrams and line drawings, monochrome photographs, extensive bibliography.

By lake and stream.
On and in freshwater.

Clegg, J., 1965. *The Freshwater Life of the British Isles* (Ed. 3) Warne (Wayside and Woodland Series), London. Guide to plants and animals of fresh water. Descriptions, biology and habits of organisms, often of species. Chapter on life in freshwater. No keys. Monochrome and colour-plates. Line drawings, selected references.

Macan, T.T., 1959. *A Guide to Freshwater Invertebrate Animals* Longmans, London. Annotated keys to freshwater invertebrates, sometimes to family, sometimes to species. Notes on parasitic forms. Line drawings.

Macan, T.T. and Worthington, E.B., 1951. *Life in Lakes and Rivers* Collins (New Naturalist), London. Freshwater biology, aquatic environments, ecology of animals and plants, fish stocks and fish ponds. Pollution. Colour and monochrome photographs; line drawings; selected references.

Mellanby, H., 1968. *Animal Life in Fresh Water* (Ed. 3) Chapman & Hall, London. Systematic guide to various groups with notes on biology. Line drawings, text references.

On the sea shore.

Barrett, J.H. and Yonge, C.M., 1958. *Collins' Pocket Guide to the Sea Shore* Collins, London. General guide to shore animals and plants: text arranged as series of keys to species. Line drawings, monochrome and colour plates. General notes on shore life; selective bibliography.

Eales, N.B., 1967. *The Littoral Fauna of the British Isles* (4th Ed.) Cambridge Univ. Press, London. An annotated systematic review of the British littoral fauna. Keys to and short descriptions of species. Line drawings, glossary, bibliography with each section, Notes on zonation of animals on shore. List of marine biological stations in Britain.

Evans, I.O. (editor) 1962. *Observers Book of Sea and Sea-shore* Warne, London. Elementary account of sea and sea-shore, tides, animals and plants. Line drawings, colour and monochrome plates.

Southward, A.J., 1965. *Life on the Sea-shore* Heinemann, London. A brief account of the shore environment; the adaptations to shore-life, the zonation and methods of studying shore organisms. Line drawings, monochrome plates.

Yonge, C.M., 1949. *The Sea Shore* Collins (New Naturalist), London. Lucid broad account of animals and plants of sea-shore detailing habits, ecology and behaviour. Line drawings, colour and monochrome photographs.

On man and beast.

James, M.T. and Harwood, R.F., 1969. *Herm's Medical Entomology* (Ed. 6), Macmillan, London. General introduction to insects and arachnids. Details of mouth-parts. Disease transmission, arthropods of medical importance, defences, allergens etc. Extensive bibliography, monochrome plates, line diagrams.

Lapage, G., 1968. *Veterinary Parasitology* (Ed. 2) Oliver and Boyd, London. A general textbook on nematodes, platyhelminthes, arthropods, protozoa etc. that affect domestic animals. Line drawings, monochrome photographs, references.

Smith, K.G.V. (Ed.) 1973. *Insects and other Arthropods of Medical Importance* B.M.N.H., London. Monographic accounts of medically important insects by various authorities. Keys to genera and sometimes to species. Vector tables. (Deals with world species). Line drawings, colour and monochrome plates.

Snow, K.R., 1974. *Insects and Disease* Routledge & Kegan Paul, London. General elementary account of insects and the diseases they carry. Line drawings, short bibliography.

Ministry of Agriculture, Fisheries and Food advisory leaflets deal with a few species on farm animals.

In house and home.

Busvine, J.R., 1966. *Insects and Hygiene* Methuen, London. Deals with the biology and control of insect pests of medical and domestic importance in Britain. Household, wood-boring, refuse, food and nuisance pests are included. Notes on insect structure, classification, physiology, ecology, & preservation; Keys for identification; notes on chemicals for control. Line drawings.

Section II

General books on the invertebrates.

Barnes, R.D., 1968. *Invertebrate Zoology* (Ed. 2) W.B. Saunders, London. Students textbook. Insects not dealt with in detail but coverage of other groups excellent. Line drawings. References chapter by chapter.

Borradaile, L.A., Eastham, L.E.S., Potts, F.A. & Saunders, J.T., 1961. *The In-*

vertebrata (4th Ed. revised by G.A. Kerkut) Cambridge Univ. Press, Cambridge. A text book dealing with all groups of invertebrates. Valuable chapter on literature of subject. Line drawings.

Buchsbaum, R., 1948. *Animals without Backbones* (Vols. I & II) Pelican Books, Harmondsworth. A readable, inexpensive, semi-technical account of the invertebrates. Text drawings and monochrome photographs. No refs.

Buchsbaum, R. & Milne, L.J., 1960. *Living Invertebrates of the World* Hamish Hamilton, London. A general account of the invertebrates excluding the insects. Living habits are emphasised and only essential details of structure are given. Line drawings; monochrome and colour photographs. Selective bibliography.

Marshall, A.J. & Williams, W.D., 1972. *Textbook of Zoology Invertebrates* (Ed. 7 of Parker & Haswell, Textbook of Zoology, Vol 1). Macmillan, London. General account of anatomy, functions, classification etc. of invertebrates. Line drawings, references.

Money, S., 1970. *The Animal Kingdom* Hamlyn, London. Elementary general introduction to animals. Colour drawings, brief glossary.

Whiteley, D., Nichols, D. & Cooke, J.A.L., 1971. *The Oxford Book of Invertebrates* Oxford Univ. Press, London. General elementary introduction to invertebrates except insects. Several species detailed from each major group. Many colour plates. Synopsis of classification, short glossary and bibliography.

General books on insects.

Chinery, M., 1973. *A Field Guide to the Insects of Britain and Northern Europe* Collins, London. General introduction to the study of insects. Essential data on structure, life-history, collecting and preservation. Concise accounts of many families and commoner species. Keys as far as families. Glossary, selected bibliography, line drawings, many fine colour pictures.

Imms, A.D., 1957. *A General Textbook of Entomology* (9th Ed. revised by O.W. Richards & R.G. Davies) Methuen, London. Essential textbook for the serious student; includes anatomy, physiology, development and classification of in-

sects. Many line drawings, extensive bibliography.

Imms, A.D., 1971. *Insect Natural History* (Ed. 3) Collins (New Naturalist), London. A learned but lucid and nontechnical account of the structure, classification and ways of life of British Insects. Line drawings, monochrome and colour plates; selective bibliography.

Imms, A.D., 1967. *Outlines of Entomology* (5th Ed. revised and corrected by O.W. Richards & R.G. Davies) Methuen, London. A concise general introduction to entomology. Line drawings; selective bibliography.

Moreton, B.D., 1950. *Guide to British Insects* Macmillan, London. Elementary introduction. Anatomy, metamorphosis, life histories etc. Key to orders. Orders described—emphasis on pest species. References, glossary, index of popular and scientific names. Line drawings, monochrome photographs.

Klots, A.B. & Klots, E.B., 1959. *Living Insects of the World* Hamish Hamilton, London. General non-technical account of insects with emphasis on life-habits. Coverage to family and sometimes subfamily level. Line drawings, monochrome and colour plates. Selective bibliography.

Wigglesworth, V.B., 1966. *The Life of Insects* Weidenfeld & Nicolson, London. General book on how insects live: adaptation to terrestrial life, movements, food, reproduction, colours, defences, senses, social organisation etc. Line drawings, photos (mainly monochrome), glossary, selected references.

Wigglesworth, V.B., 1974. *Insect Physiology* (7th Ed.), Chapman & Hall, London. Elementary account.

Parasitic insects.

Askew, R.R., 1971. *Parasitic Insects,* Heinemann, London. Review of parasitic species on both vertebrates and insects. Line drawings. Detailed bibliography.

General books on insect orders
(some contain taxonomic material).

Odonata.

Corbet, P.S., Longfield, C. & Moore, N.W., 1960. *Dragonflies,* Collins (New Naturalist) London. British dragonflies; their history, distribution, life-histories, behaviour, dispersal, seasonal regulation and relationships with animals and man. Key to larvae, distribution maps, line drawings, colour and monochrome photographs, references and glossary.

Orthoptera, Dictyoptera, Phasmida.

Ragge, D. R., 1965. *Grasshoppers, Crickets and Cockroaches of the British Isles,* Warne (Wayside & Woodland Series), London. Orthoptera, Dictyoptera and Phasmida native to or introduced into Britain. Keys and descriptions of species. Habitats, habits and life histories described; notes on ecology and timing. Collecting, preserving and rearing. Song diagrams. (N.B. record of songs can be bought). Distribution and explanation of vice-county scheme. Line drawings, colour plates. Check list with common names. Glossary, selective bibliography.

Hemiptera (Heteroptera)

Southwood, T.R.E. & Leston, D., 1959. *Land and Water Bugs of the British Isles,* Warne (Wayside and Woodland Series), London. Account of British Hemiptera-Heteroptera. Keys to and descriptions of species. Life-histories, habits, distribution. Line drawings, monochrome photographs, colour plates. Notes on collecting, preservation, dissection of genitalia. Glossary, detailed bibliography in several sections.

Lepidoptera

Beirne, B.P., 1952. *British Pyralid and Plume Moths,* Warne (Wayside and Woodland Series), London. Account of British species of Pyralidae, Pterophoridae and Orneodidae. Keys and descriptions of species, habits etc. Line drawings, colour photographs, maps. Glossary.

Higgins, L.G., & Riley, N.D., 1970. *A Field Guide to the Butterflies of Britain and Europe,* Collins, London. Handbook to European species. Species described with timing, habitat, distribution etc. Line diagrams, many colour plates. Checklist, glossary, distribution maps, selective bibliography.

South, R., 1961. *The Moths of the British Isles* Series I and II (New edition, edited by H.M. Edelsten, D.S. Fletcher and R. J. Collins), Warne (Wayside and Woodland Series), London. Macrolepidoptera of Britain. Brief descriptions and notes on distribution, habits, foodplants and timing of species. Drawing of larvae,

pupae, and food-plants. Colour plates of adults. No keys.

Stokoe, W.J. & Stovin, G.H.T., 1944. *The Caterpillars of the British Butterflies,* Warne (Wayside and Woodland Series), London. Eggs, larvae and pupae of all British butterflies described. Habits, timing and food plants. Illustrations of food plants and species of butterflies associated listed. Colour and monochrome plates.

Stokoe, W.J. & Stovin, G.H.T., 1948. *The Caterpillars of the British Moths* (2 vols), Warne (Wayside and Woodland Series), London. Eggs, larvae and pupae of British Macro-lepidoptera described. Habits, food plants etc. Colour and monochrome plates.

Trichoptera.

Hickin, N. E., 1967. *Caddis Larvae, Larvae of the British Trichoptera,* Hutchinson, London. Monograph on caddis larvae. Evolution, lifecycle, collecting, preserving, biology, importance in natural economy. Detailed accounts of separate species with keys. Line drawings, detailed bibliography.

Diptera.

Colyer, C.N. and Hammond, C.O., 1968. *Flies of the British Isles,* Warne (Wayside and Woodland Series), London. General account of biology of British flies with particular reference to more common species. Colour and monochrome plates. References.

Oldroyd, H., 1964. *The Natural History of Flies,* Weidenfeld & Nicolson, London. Monograph on biology of flies. Monochrome plates. Many references.

Hymenoptera.

Free, J.B. & Butler, C.G., 1959. *Bumblebees,* Collins (New Naturalist), London. Monograph on Bumblebees: biology, behaviour, colonies, economic importance. Notes on collecting and distribution. Keys to *Bombus* and *Psithyrus* species. Line drawings, 1 colour plate, 24 monochrome plates. Bibliography.

Step, E., 1946. *Bees, Wasps, Ants and Allied Insects of the British Isles,* Warne (Wayside and Woodland Series), London. General account of Hymenoptera. Conspicuous species considered at length but book not intended for identification to species. Monochrome and colour plates, line drawings. Short glossary and bibliography.

General books on other arthropod groups (some contain taxonomic material).

Bristowe, W.S., 1958. *The World of Spiders*, Collins (New Naturalist), London. An account of British spiders, family by family. Chapters on history and folklore, structure and classification etc. Notes on collecting and preservation. Check list of British species with regional distribution. Bibliography. Line drawings, monochrome and colour photographs.

Eason, E.H., 1964. *Centipedes of the British Isles*, Structure, biology, distribution, habitat, classification and nomenclature of British species. Keys and descriptions of species. Notes on collection and preservation. Line drawings, 1 colour plate and several monochrome plates. Glossary, references.

Sutton, S.L., 1972. *Woodlice*, Ginn & Co. Ltd., London. A general account of the structure, physiology, ecology and distribution of British woodlice. Check list, keys to and notes on individual species. Line drawings, colour plates. Notes on collecting, rearing etc. Experiments with woodlice, bibliography.

Section III

Collection and preservation.

Many of the books listed above and below contain notes on the collection and preservation of invertebrates. The following may also be noted.

British Museum (Natural History) Handbook. *Instructions for collectors No. 9A. Invertebrate animals other than insects.* 2nd Ed. 1954.

Oldroyd, H. 1970. *Collecting, Preserving and Studying Insects.* Hutchinson, London.

Identification.

There is a vast literature on the identification of British invertebrates much of which is out of print and out of date. The following source book is invaluable to any person interested in finding what is available for dealing with various groups.

Kerrich, G.J., Meikle, R.D. & Tebble, N., 1967. *Bibliography of Key Works for the Identification of the British Fauna and Flora* (Ed. 3), Systematics Association, London, A guide to the literature needed to identify British species.

Guidance on the systematic arrangement of animals in general can be had from:

Rothschild, Lord, 1965. *A Classification of Living Animals* (Ed. 2), Longmans, London.

The following sources are chosen because they are available, authoritative and, for the most part, up-to-date. Note that animals of "popular" groups can often be identified by means of the books listed in Section II.

Land invertebrates in general, but not insects.

Cloudsley-Thompson, J.L. & Sankey, J., 1961. *Land Invertebrates*, Methuen, London. Concise general guide to British land invertebrates excluding insects. Species identifications except in difficult groups. No keys. Line drawings, selective bibliography.

Fresh water species in detail.

Freshwater Biological Association. Scientific Publications Issued in parts:

5. *A Key to the British Species of Freshwater Cladocera*, D.J. Scourfield & J.P. Harding, (3rd Ed.); 1966.

8. *Keys to the British Species of Aquatic Megaloptera and Neuroptera*, D.E. Kimmins, (2nd Ed.); 1962

13. *A Key to the British Fresh- and Brackish-Water Gastropods*, T.T. Macan, (3rd ed.); 1969.

14. *A Key to the British Freshwater Leeches*, K.H. Mann, (2nd ed.); 1964.

15. *A Revised Key to the Adults of the British Species of Ephemeroptera*, D.E. Kimmins, (2nd revised ed.); 1972.

16. *A Revised Key to the British Water Bugs (Hemiptera-Heteroptera)*, T.T. Macan, (2nd ed.); 1965.

17. *A Key to the Adults and Nymphs of the British Stoneflies (Plecoptera)*, H.B.N. Hynes, (2nd revised ed.); 1967.

18. *A Key to the British Freshwater Cyclopid and Calanoid Copepods*, J.P. Harding & W.A. Smith; 1960.

19. *A Key to the British Species of Crustacea: Malacostraca occurring in Fresh Water*, H.B.N. Hynes, T.T. Macan & W.D. Williams; 1960.

20. *A Key to the Nymphs of British Species of Ephemeroptera*, T.T. Macan. (2nd revised ed.); 1970.

22. *A Guide for the Identification of British Aquatic Oligochaeta*, R.O. Brinkhurst (2nd ed.); 1971.

23. *A Key to the British Species of Freshwater Triclads*, T.B. Reynoldson: 1967.

24. *A Key to the British Species of Simuliidae (Diptera) in the Larval, Pupal and Adult Stages*, Lewis Davies; 1968.

26. *A Key to the Larvae, Pupae and Adults of the British Species of Elminthidae*, D.G. Holland; 1972.

28. *A Key to the Adults of the British Trichoptera*, T.T. Macan & C.J. Worthington; 1973.

Insects in general.

Chu, H.F., 1949. *How to Know the Immature Insects*. Brown, Iowa. Keys and line drawings enabling most immature insects to be run down to family. Selected references. Glossary.

Insects in detail.

Collingwood, C.A., 1964. *The Identification and Distribution of British Ants. 1. A Revised Key to the Species found in Britain*. Transactions of the Society for British Entomology, 16, 93 – 114.

Ford, L.T., 1949. *A Guide to the smaller British Lepidoptera*, South London Entomological and Natural History Society, London. An annotated catalogue of smaller species giving details of food plants. No illustrations, no keys, no glossary, but helps identification by listing species associated with each food plant.

Joy, N.H., 1932. *A Practical Handbook of British Beetles*, Vols I and II, Witherby, London. Keys for identification of species in Vol I; line drawings of many of species in Vol 2. Notes on collecting and preservation and glossary.

Meyrick, E., 1928. *A Revised Handbook to British Lepidoptera*, Watkins & Doncaster, London. Technical handbook to British species. Keys, few line drawings, no plates. Brief glossary. List of food plants and of their common names.

Royal Entomological Society of London; *Handbooks for the identification of British Insects* Vols I – XI. These handbooks are published as parts are completed. They aim eventually to provide illustrated annotated keys to the whole of the British insect fauna

in ten volumes. The eleventh volume is a revised (2nd) edition of Kloet & Hincks Check List of British Insects. Parts so far available are:

Vol I

Pt.2. *Thysanura & Diplura*, M.J. Delany; 1954.

Pt.5. *Dermaptera & Orthoptera*, W.D. Hinks; 1956 (2nd ed.).

Pt.6. *Plecoptera*, D.E. Kimmins; 1950.

Pt.7. *Psocoptera*, T.R. New; 1974.

Pt.9. *Ephemeroptera*, D.E. Kimmins; 1950.

Pt.10. *Odonata*, F.C. Fraser; 1956 (2nd ed.).

Pts.12 to 13. *Mecoptera, Megaloptera, Neuroptera*, F.C. Fraser; 1959.

Pt.16. *Siphonaptera*, F.G.A.M. Smit; 1957.

Vol. II Hemiptera.

Pt. 2(a) *Hemiptera-Homoptera: Cicadomorpha* (pt.), W.J. Le Quesne; 1965.

Pt. 2(b) *Hemiptera-Homoptera: Cicadomorpha* (contd.), W.J. Le Quesne: 1969.

Pt. 3. *Hemiptera-Homoptera: Fulgoromorpha*, W.J. Le Quesne; 1960.

Vol. IV Coleoptera (i)

Pt. 1. *Introduction and Key to Families*, R.A. Crowson; 1956.

Pt. 2. *Carabidae*, C.H. Lindroth; 1974.

Pt. 3. *Hydradephaga*, F. Balfour-Browne; 1953.

Pt. 6(a) *Clambidae*, C. Johnson; 1966.

Pt. 8(a) *Staphylinidae* (pt.) C.E. Tottenham; 1954.

Pt. 9. *Pselaphidae*, E.J. Pearce; 1957.

Pt. 10. *Sphaeritidae & Histeridae*, D.G.H. Halstead; 1963.

Vol. V Coleoptera (ii)

Pt. 5(b) *Phalacridae*, R.T. Thompson; 1958.

Pt. 7. *Coccinellidae & Sphindidae*, R.D. Pope; 1953.

Pt. 9. *Lagriidae to Meloidae*, F.D. Buck; 1954.

Pt. 11. *Scarabaeoidea*, E.B. Britton; 1956.

Pt. 12. *Cerambycidae*, E.A.J. Duffy; 1952.

Pt. 15. *Scolytidae & Platypodidae*, E.A.J. Duffy; 1953.

Vol. VI Hymenoptera (i)

Pt. 1. *Introduction and Key to Families*, O.W. Richards; 1956.

Pt. 2(a) *Symphyta* (pt.), R.B. Benson; 1951.

Pt. 2(b) *Symphyta* (contd.), R.B. Benson; 1952.
Pt. 2(c) *Symphyta* (concl.), R.B. Benson; 1958.

Vol. VII Hymenoptera (ii)
Pt. 2(ai) *Ichneumonoidea* (pt.), J.F. Perkins; 1959.
Pt. 2(aii) *Ichneumonoidea* (contd.), J.F. Perkins; 1960.

Vol. VIII Hymenoptera (iii)
Pt. 1 (a) *Cynipoidea* (pt.), R.D. Eady & J. Quinlan; 1963.
Pt. 2 (a) *Chalcidoidea* (pt.), Ch. Ferrière & G.J. Kerrich; 1958.
Pt. 2 (b) *Chalcidoidea* (cont.), R.R. Askew; 1968.
Pt. 3 (dii) *Proctotrupoidea* (pt.), G.E.J. Nixon; 1967.

Vol. IX Diptera (i)
Pt. 1. *Introduction and Key to Families,* H. Oldroyd; 1970 (3rd ed.).
Pt. 2. *Nematocera* (pt.), R.L. Coe, P. Freeman, P.F. Mattingly; 1950.
Pt. 4. *Tabanoidea & Asiloidea,* H. Oldroyd; 1969.

Vol. X Diptera (ii)
Pt. 1.   *Syrphidae,* R.L. Coe; 1953.
Pt. 2 (ai) *Lonchopteridae,* K.G.V. Smith; 1969.
Pt. 2 (c) *Pipunculidae,* R.L. Coe; 1966.
Pt. 3 (a) *Conopidae,* K.G.V. Smith; 1969.
Pt. 4 (a) *Cyclorrhapha,* (*Tachinidae, Calliphoridae),* F.I. van Emden; 1954.
Pt. 4 (b) *Cyclorrhapha* (*Muscidae*), E.C.M. d'Assis-Fonseca; 1968.
Pt. 5 (g) *Agromyzidae,* K.A. Spencer; 1972.

Vol. XI Kloet & Hincks Check List (ed. 2).
Pt. 1. *Small orders,* Hemiptera; 1964.
Pt. 2. *Lepidoptera;* 1972.

Kloet, G.S. & Hincks, W.D., 1945. *A Check List of British Insects.* The only complete list of British Insects. Second edition now being produced in parts. (See also Handbooks for the Identification of British Insects; Royal Entomological Society of London).

Other arthropods.

Arthur, D.R., 1963. *British Ticks,* Butterworth, London. Technical manual to British ticks. Keys to all stages, descriptions, distributions, hosts etc. of species. Line drawings, bibliography.

Balogh, J., 1972. *The Oribatid Genera of the World,* Budapest. Technical manual to oribatid mites. Keys to genera; catalogue of genera and type-species; frequent synonyms. Line drawings of many species.

Locket, G.H. & Millidge, A.F., 1951, 53. *British Spiders,* Vols. I and II. Ray Society, London. Taxonomic account of British Spiders. Keys to families and genera. Species described and illustrated. Notes on collecting; structure, check list, refs.

Locket, G.H., Millidge, A.F. & Merrett, P. 1974. *British Spiders,* Vol III, Ray Society, London. Additions to species dealt with in Vols I and II, corrections etc. Distribution maps by vice-county of 612 species, check list. Line drawings, bibliography.

Molluscs.

Ellis, A.E., 1926. (re-issued 1969). *British Snails,* Clarendon Press, Oxford. Guide to non-marine gastropods of Great Britain and Ireland. Descriptions, check list, notes on structure, classification etc. Pliocene to Recent species all included. Monochrome photographs, line drawings, references.

Mc Millan, N.F., 1968. *British Shells,* Warne (Wayside and Woodland Series), London. British molluscs (including slugs and nudibranchs). Land, freshwater and marine species considered. Descriptions mainly of shells. Key to slugs. Line drawings, colour plates. Check list. Notes on collecting and economic uses. Glossary. Annotated selective bibliography.

Tebble, N., 1966. *British Bivalve Seashells,* British Museum (Natural History), London. Technical handbook to seashells. Keys for identification of species; description and notes on biology, distribution and habits of species. Glossary, line drawings, colour & monochrome plates.

Miscellaneous groups.

The Linnean Society of London, *Synopses of the British Fauna,* Linnean Society, London.
Parts issued from time to time as follows:
2. Harrison, J.R., 1944, *Caprellidae (Amphipoda).*
3. Reid, D.M., 1944, *Gammaridae.*
4. Ellis, A.E., 1946, *Fresh water bivalves (Corbicula, Sphaerium etc.).*

5. Ellis, A.E., 1947, *Fresh water bivalves (Unionacea)*.
6. Gerard, B.M., 1964, *Lumbricidae*.
7. Reid, D.M., 1947, *Talitridae (Amphipoda)*.
8. Quick, H.E., 1949, *Slugs*.
9. Edney, E.B., 1954, *British Woodlice*.
10. Evans, G.O. & Browning, E., 1954, *British Pseudoscorpions*.
11. Blower, J.G., 1958, *British Millipedes (Diplopoda)*.
12. Stephen, A.C., 1960, *British Echiurids, Sipunculids & Priapulids*.

New Series.

1. Millar, R.H., 1970, *British Ascidians*.
2. Graham, A., 1971, *British Prosobranchs*.
3. Naylor, E., 1972, *British Marine Isopods*.
4. Sankey, J.H.P. & Savory, T.H., 1974, *British Harvestmen*.
5. King, P.E., 1974, *British Sea-spiders*.

Glossary

N.B. Where terms have more than one meaning, usually only that relating to invertebrate zoology is included.

Abdomen paunch, hindermost region of insect body.

Abdominal relating to the abdomen.

Aberration deviation from type, e.g. in colour or structure.

Acaricide poison used for killing mites.

Aculeate having a sting (Hymenoptera).

Acuminate narrowing to a point.

Adaptation modification for particular purpose.

Adult fully mature individual.

Aedeagus intromittent organ and its immediate surroundings of male genitalia of insect (structure much used for identifying species).

Air-sacs sacs in the tracheal system of many insects. Their principle function is to facilitate air-exchange. Resting insects may often be seen to make rapid breathing movements during which the air in the air-sacs is changed. Passage of air through the small tracheae (air-tubes) is thereafter by diffusion only.

Albinism abnormal condition due to absence of pigment in skin. Compare Melanism.

Allergic over-sensitivity, e.g. to irritant.

Alternation of generations See Generation alternation.

Ametabola insects which undergo no metamorphosis (q.v.) during development.

Amphibious living both on land and water.

Anal relating to anus, also sometimes relating to hind part, e.g. anal lobe of wing in insect.

Anal horn a prominent often curved process on the tip of the abdomen of hawk-moth caterpillars (e.g. no. 102).

Anatomy study of form and structure of an organism.

Antennae the feelers on the heads of most insects, centipedes etc. (one pair) and of crustaceans (two pairs). They are furnished with organs of touch and often of smell. Their shape varies greatly amongst different animals. (The tentacles of snails are sometimes called antennae.)

Anterior foremost, e.g. anterior wings (c.f. Posterior).

Anus hind opening of gut.

Apical relating to the part of appendage furthest from the body (c.f. Basal).

Appendage limb or other articulated process of body.

Arctic relating to area around N. Pole.

Army-worms term used for caterpillars of certain moths and also for larvae of certain flies (e.g. no. 416) which form and move together in columns or large masses.

Author in biological nomenclature a person who first describes and names a species (q.v.) or other taxon (q.v.) in a publication. The author's name quoted after the name of species and followed by the date defines the species absolutely, e.g. *Musca domestica* Linnaeus, 1758 is the 'house-fly'. The name of an author is often abbreviated conventionally e.g. L. = Linnaeus and if an animal is transferred to another genus his name is placed in brackets e.g. *Pieris brassicae* (L., 1758). For well known animals the author's name and the date are usually omitted.

Autotomy self-amputation of limb or other part of body usually to escape danger or when cast part is no longer needed. It occurs in crustaceans, harvestmen, brittle-stars etc. Often the lost parts can be re-generated (see Regeneration).

Autotrophic describes organisms which build up organic materials from inorganic ones, e.g.s. green plants and certain bacteria c.f. Heterotrophic.

Basal relating to part of appendage nearest to the body or to part of body structure nearest to centre of body.

Biocide chemical poisonous to life (e.g. insecticide, fungicide) used to control pests etc.

Biology study of living organisms.

Biological control control of animals or plants by use of natural enemies such as parasites & predators.

Bioluminescence light production by living organisms (e.g. nos. 112, 388).

Biotope territory with defined conditions e.g. soil, moisture, atmosphere etc. which accommodates a certain type of animal and plant community.

Botany study of plants.

Brood-care care of the young by the parent. Though more characteristic of birds and mammals, brood-care occurs in some in-vertebrates – especially amongst the in-sects. It is most developed in social forms, e.g. ants, bees, wasps and termites but also occurs in certain solitary forms e.g. ear-wigs (no. 241) and sexton beetles (no. 253).

Budding a form of asexual reproduction in which part of an organism separates off to form a new individual. Many unicellular organisms bud off daughter cells and mul-ticellular forms such as hydroids and jelly-fish may bud off complete new individuals.

Camouflage disguise, colouration etc. to merge with surroundings as a protection.

Cannibalism eating of animals by members of own species. Cannibalism occurs in many predators e.g. spiders and mantises but may also occur amongst the larvae of some Lepidoptera and amongst young snails.

Cantharidin a poisonous substance occur-ring in certain beetles e.g. some members of the family *Meloidae* (no. 113). It has been used in medicine.

Carnivore flesh-eater, e.g. predator.

Caste specialized type of individual found in social insects. e.g. in honey-bee the castes are queens, workers and drones.

Cephalothorax combined head and thorax as in arachnids. Preferred term is prosoma (see p. 177).

Cerci a pair of appendages on the tip of an insect's abdomen.

Chaetae bristles such as occur in many annelid worms.

Chelicerae the most anteriorly-placed appendages of Arachnida and Xiphosura. They vary greatly in form. In spiders they are fang-like and have poison canals so that prey can be bitten and paralysed or killed (see p. 177).

Chemical control control of animal pests or of undesirable plants by use of chemicals (cf. Biological control).

Chemo-receptor receptive organs for senses of smell and taste. In invertebrates these senses are not so well differentiated as in vertebrates; they have been most in-vestigated in insects. The sense of smell operates at a distance and may guide the animal to food, the opposite sex or to an egg-laying site. It is most often located on the antennae. The sense of taste is a contact sense; it helps the animal to test its food etc. and is often located on the mouth-parts or legs.

Chitin a cellulose-like substance found in the cuticle of arthropods, in molluscs, in some worms and in fungal cell-walls. It is resistant to chemicals and mechanical damage.

Chromosomes small elongated structures found within the nuclei of cells. Their number is constant for each species. They carry the genes or heredity factors. In the science of genetics the study of chromo-somes is important. Chromosomes usually only become readily visible during cell division; certain flies (e.g. no. 415) have particularly large chromosomes in their salivary glands and these are much used for study.

Cilia hair-like processes from cells. They occur in large numbers, show constant motion and beat in a defined direction. Found in most groups of animals they form the main organs of locomotion in ciliates and planarians (no. 358). In mammals the trachea and bronchi are lined with ciliated cells.

Claspers paired appendages of male insects used for holding the female during copu-lation.

Clavate clubbed.

Clitellum saddle-like glandular region of body in mature oligochaetes and leeches (indistinct in latter). It secretes the cocoon in which the eggs are deposited.

Clypeus head plate of insect; the region immediately above the upper lip.

Cocoon protective sheath. Larvae of Lepidoptera often spin cocoons of silk around themselves before they pupate. Spiders spin egg-cocoons and earthworms secrete tough bag-like egg-cocoons from the glands of the clitellum.

Colonial animals these can be divided into two sorts

(a) social forms such as ants and bees. These live in colonies of many individuals and the work of the whole community is shared amongst them

(b) forms such as corals and sea-firs. Here large closely packed communities are built up by asexual reproduction and in many instances the separate 'individuals' are connected by living tissues.

Colour changing some animals can change colour during their life-time. This often has a protective function as they come to resemble their backgrounds, e.g. grasshoppers can slowly adjust their colours and the crab-spider *Misumena vatia* (no. 44) can change from white to yellow or vice versa according to the colour of the flower it inhabits. Cuttle-fish and chamaeleons can change colour rapidly.

Colour variants many insect species occur in several forms with different colour markings. Such occurrences are common in Lepidoptera, Hemiptera, Hymenoptera and Orthoptera. The different markings may be hereditarily determined or may be due to environmental factors (temperature, light, humidity etc.) in which case they are not inherited.

Commensalism the relationship between two un-related organisms that live together but are not dependent on each other. Both may however gain some benefit from the relationship. c.f. Parasitism and Symbiosis.

Compound eyes eyes formed of many separate elements or ommatidia. Compound eyes occur in most insects, in crustaceans and in king-crabs.

Contact poison a pest-control chemical which is absorbed and acts through the skin or cuticle of an animal.

Copulation sexual intercourse.

Cosmopolitan having an almost worldwide distribution.

Coxa the basal segment of an insect leg.

Cross-breeding breeding between individuals of different species or sub-species. The off-spring are hybrids (q.v.).

Cuckoo-spit drop of foam produced by the larva of the bug *Philaenus spumarius* (no. 117).

Cuticle the non-cellular outermost covering of many animals. In arthropods, the cuticle is complicated and varies in thickness and texture in different parts of the animal. In some parts it may be thickened and hardened to form an armour-like

exoskeleton; over joints it is thin and flexible. In insects it is composed of chitin and tanned protein (sclerotin); in crustaceans it is hardened by impregnation with calcium salts. The cuticle has to be cast off periodically to allow of growth (see Ecdysis).

Cyst bladder or sac. The resistant resting stages of single-celled animals and trematodes. The spherical egg-laden dead bodies of certain plant-parasitic nematodes (e.g. no. 66) are also called cysts.

D.D.T. abbreviation for persistent contact insecticide dichlorodiphenyl trichloroethane.

Detritus organic matter derived from the break-down of dead plants and animals which is still in the form of particles (c.f. Förna and Humus).

Diapause a resting stage during the life of an organism, characterised by a cessation of feeding and a reduced metabolic rate. Animals in diapause often show enhanced resistance to adverse conditions, e.g. low temperature and drought. In insects diapause is in some species associated with a particular stage of the life-cycle but in other species may occur at various stages. Overwintering diapause (hibernation) may commence as a response to shortening day-length in autumn, i.e. it is not a direct response to adverse conditions, and its termination may be preceded by a period of chilling in the spring. Diapause enables an organism to survive in areas where the environment is not suitable for active life throughout the year.

Dimorphism when a species has two different forms it exhibits dimorphism. Sexual dimorphism is almost universal and the sexes can differ to varying degrees. As well as differences associated with the reproductive organs it can manifest itself as differences in colour, size, degree of development of sensory organs, wings, mandibles etc. Seasonal dimorphism also occurs, e.g. many birds have different winter and summer plumage, and some butterflies and bugs have winter and summer forms.

Distal the part of a limb etc. furthest from the body.

Disease vectors many animals, particularly insects and mites with mouth parts adapted for piercing and sucking spread diseases to animals and plants. Some examples are:

Disease	Vector	Cause of disease
Bubonic plague	Flea	bacterium
Typhus	human louse	rickettsia
Yellow fever	mosquito	virus

Malaria	mosquito	sporozoan
Sleeping sickness	tsetse fly	flagellate
Potato leaf roll	aphid	virus
Turnip yellow mosaic	flea beetle	virus

Dorsal relating to the back, i.e. side usually held uppermost.

Drone male of bee.

Ecdysis moulting or casting off of cuticle. Arthropods have a more or less rigid exoskeleton and cast it off periodically to enable growth to take place. Before moulting the inner layers of the cuticle are dissolved and re-absorbed and at the same time a new cuticle starts to form beneath the old one. After the remains of the old cuticle have been discarded the new cuticle stretches before it becomes hard. Moulting is controlled by hormones and the number of moults varies from species to species. Nematodes moult also.

Ecology the study of living things in relation to each other and to the environment they live in.

Ecosystem the complex of plant, animal and environment in a locality. Materials and energy circulate in an ecosystem and are taken from and passed out to the surrounding district. The organisms within the system can be classified into producers, consumers and destroyers–the latter break up organic substances into inorganic ones (see Food chain).

Ectoparasite see Parasite.

Elytra the hardened wing-cases, i.e. the modified fore-wings of beetles.

Embryo foetus, young unborn animal within its mother or within an egg-shell.

Embryology study of foetus development.

Endemic a species is endemic to a district if it is native to it.

Endoparasite see Parasite.

Endopterygota see page 178.

Entomology the study of insects.

Environment the medium in which an organism lives. It includes the physical environment, i.e. air, water, soil, temperature etc. and the biological environment, i.e. other organisms of the same and different species.

Enzymes proteins found in living organisms which permit or accelerate specific chemical processes. Enzymes participate in almost all biochemical reactions e.g. those in the digestive juices break the complex molecules of the food into simple ones.

Ephippium a sheath round the resting eggs of water-fleas. It is formed from the thickened brood-sac walls of the female.

Ethology the study of animal behaviour.

Evolution the gradual change of the characteristics of organisms during the course of many generations by means of which they acquire specialised structures and diverge into separate species.

Exarate pupa an insect pupa in which the wings, limbs etc. are freely moveable, e.g. in Coleoptera (most) and Hymenoptera.

Exopterygota see page 178.

Exuvium the remains of the cuticle cast off at ecdysis.

Facet the surface of one of the units making up the compound eye of an arthropod.

Fauna the animal world; the animal population of a locality.

Femur the largest segment of the insect leg.

Flagellum whip. Thin thread-like structure present in certain lower animals and often used as a locomotory organ. Flagella are longer than cilia (q.v.) and there is only one or a small number per cell. The same name is used for the outer part of the insect antenna.

Flora the plant world; the plant population of an area.

Flight muscles insects have two sets of flight muscles, the direct ones and the indirect ones. The direct muscles are attached to the wing base and wall of the thorax and in the more primitive insects such as dragonflies and grasshoppers raise and lower the wings and are the main flight muscles. In higher insects such as bees and wasps, however, the indirect muscles provide the power for flight. There are two sets of these, running horizontally-lengthwise and vertically and their alternate action elevates and depresses the thorax capsule causing the wings to fall and rise correspondingly. In these insects the direct muscles twist the wings during each beat.

Förna remains of dead animals and plants persisting in the upper layers of the soil. Förna particles consist of undecomposed matter. (c.f. Detritus and Humus.)

Fossil remains of a prehistoric organism. Usually only the hard parts are preserved and are partially or wholly replaced by mineral materials. Foot-prints of prehistoric animals can also be preserved as fossils.

Fossorial adapted for digging.

Food chain the course taken by food materials through a series of different species in a natural community, starting with the producers and ending with the destroyers, e.g.

(Continues overleaf)

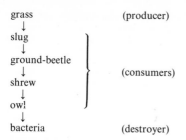

grass   (producer)

slug

ground-beetle   (consumers)

shrew

owl

bacteria   (destroyer)

Fright-colouration/Fright pattern certain insects display striking colours or patterns often in combination with sudden movements to scare would-be enemies. Some of the most effective are the 'eye-spots' seen in forms such as the peacock butterfly (no. 11) the emperor moth (no. 167) and the larvae of the elephant hawkmoth (no. 102) and pussmoth (no. 160).

Functional polymorphism see polymorphism.

Gall an abnormal growth on a plant produced as a response to mechanical or chemical irritation of its tissues by some organism. Galls often appear as localised swellings and may be characteristic for the particular organism concerned. Insects, mites, nematodes, bacteria and fungi can all cause gall formation. Amongst the insects aphids, coccids, psyllids, and the larvae of gall- and leaf-wasps, weevils, gall-midges, trypetids and certain lepidoptera are gall formers. Oaks in particular show many kinds of insect gall (nos. 208, 209).

Ganglion a cluster of nerve cells.

Gene a single hereditary factor (see Chromosomes).

Generation a complete brood from egg-laying of parent to that of offspring.

Generation alternation certain lower animals show an alternation of sexually and asexually reproducing generations, e.g. a medusa or jelly-fish produces fertilised eggs. These develop into polyps which bud off a succession of free-swimming medusae. Aphids also often have a sexual generation which alternates with a succession of asexual generations.

Genetics study of heredity.

Genitalia the sex organs. In many closely related and very similar species of insects the genitalia differ considerably and afford reliable characteristics for identification of the separate species.

Genus a group of closely related species. All species within a genus are given the same first scientific name, e.g. *Pieris brassicae* and *Pieris napi* are species of the genus *Pieris*.

Gregarisation massing together of individuals of a species.

Gills respiratory organs of aquatic animals (fish, molluscs, crabs etc.).

Gynandromorph an individual which abnormally exhibits a mixture of male and female characteristics. This condition is particularly noticeable in species with considerable sexual dimorphism (c.f. Hermaphrodite).

Haltere drum-stick like organs on thorax of dipterous flies. These represent highly modified hind-wings which by virtue of their gyroscopic action during flight give the insect a sense of equilibrium.

Haemoglobin the red respiratory pigment present in the blood of vertebrates, some crustacea, many annelids and a few insects such as the 'blood-worms' (larvae of *Chironomus* spp. no. 347).

Haemolymph the tissue fluids or blood of invertebrate animals such as arthropods and molluscs which have an open blood system. Here the 'blood' fills the body cavity and, bathing the tissues directly, is not differentiated from lymph as in vertebrates.

Hemimetabola another name for Exopterygota (see p. 178).

Hermaphrodite an individual with dual sexuality. This condition occurs in many groups of invertebrates, e.g. mussels, pulmonate snails, oligochaetes, leeches, sea anemones and ctenophores. Sometimes the animal alternates between the male and female condition e.g. the prawn *Pandalus borealis* is male at first but later changes to female. (c.f. Gynandromorph).

Heterotrophic description of an organism which must obtain its nourishment in the form of complex organic substances. Animals are heterotrophic whereas green plants and some bacteria are autotrophic (q.v.).

Histology the study of the microscopic structure of organic tissues and organs.

Holometabola another name for Endopterygota (see p. 178).

Holotype see Type.

Homologous organs etc. of an animal are homologous if they have a similar origin.

Homonym if two or more species or animal groups are given the same name their names become homonyms. The first animal to be named then retains the name and the others are renamed.

Honey-dew a sugary, liquid excrement from aphids and other plant bugs. It forms an important food for ants and is also sought by bees, wasps and some flies.

Host-specific an animal is host-specific when it is associated with one species of plant or animal only.

Host-alternation this occurs when animals regularly change from one type of host to another and back again. E.g. many aphids have herbaceous summer hosts but spend the winter on trees and bushes.

Humeral relating to the shoulder.

Humus decomposed organic material of soil. Degradation has reached the point where organic structure can no longer be made out. Brown earths (mull soils) raw humus (mor soils) and peats have progressively higher humus contents.

Hybrid an individual resulting from the crossing of two different species or sub-species of animals or plants. Hybrids between different species (e.g. the mule) are usually sterile.

Hypermetamorphosis a phenomenon found in many parasitic insects where, in addition to the usual metamorphosis from larva to adult, there is a change in the form of the larva between one instar and the next.

Hyperparasite see Parasitoid.

Imago the adult of an insect.

Immigrant a species which enters an area from the outside. Many flying species of insects immigrate to Britain occasionally or regularly, e.g. the death's head hawk-moth (no. 156t) and the painted lady butterfly (no. 9). In certain years great numbers of some species, e.g. the large white (no. 53) arrive from the Continent.

Industrial melanism see Melanism.

Inquiline an animal which lives in the nest, mine or gall of another species but has no apparent effect on it. Inquilines often feed on refuse or gall tissues.

Insecticide a poisonous chemical used for controlling insects.

Insect pathogens organisms such as viruses, bacteria, sporozoa and fungi which cause diseases in insects.

Instinct innate behaviour; a system of in-built activities which all contribute to an organism's characteristic behaviour pattern. The separate activities often appear as responses to certain stimuli, are of a stereotyped nature and do not have to be learned, e.g. certain parasitic wasps will attempt to lay eggs in any small roundish object which roughly resembles the eggs they usually parasitize.

Instar each separate stage of an arthropod's life between moults is an instar.

Integument another name for cuticle (q.v.).

Johnston's organ a sensory organ lying within the second antennal joint of most insects. It varies in function in different insects. Frequently it registers movements of the distal part of the antennae; in the honey-bee it measures the speed of flight and in male mosquitoes it acts as an organ of hearing picking up the weak vibrations caused by the wing-beats of the females.

Juvenile hormone a natural chemical produced within the body of an immature insect which prevents it from undergoing metamorphosis into the adult. Recently attempts have been made to use it in pest control as insects subjected to it do not mature to reproduce.

Labium the lower lip of an insect. It consists of the pair of appendages of the hindmost head segment fused together on the mid-line. It is subject to great variation amongst the different groups of insects.

Labrum the upper lip of an insect. It consists of a plate which articulates with the front margin of the head capsule.

Larva the young of an insect; first instar of a mite; immature stage of aquatic mollusc, polychaete worm etc. Usually larvae are free living and can fend for themselves but differ fundamentally in structure from the adults. The larvae of exopterygote insects are often called nymphs.

Larviparous in some insects the eggs hatch within the body of the female so that larvae are born directly. These species are said to be larviparous.

Lateral relating to the sides.

Limnic living in fresh-water.

Limnology study of fresh-water organisms and their environment.

Light production occurs in various sorts of animals, e.g. in the unicell *Noctiluca* and in certain polychaete worms, geophilid centi-pedes, crustaceans and fishes. Amongst insects it occurs in glow-worms and certain other beetles, in some spring-tails and fly larvae and occasionally elsewhere. The female of the glow-worm sends out such a strong light that it can be seen clearly from several metres and the male locates her by means of it. The light is produced by the oxidation of a substance 'luciferin' in the presence of the enzyme (q.v.) 'luciferase'.

Mandibles in arthropods except arachnids, king-crabs and a few minor groups the mandibles are mouth-parts formed from the appendages of the segment immediately behind the mouth. They are often adapted for biting or holding but may differ greatly between different species. The 'antlers' of the male stag-beetle (no. 190) are the highly modified mandibles.

Marine relating to the sea.

Maxillae the second pair of mouth parts of mandibulate arthropods. In insects they lie between the mandibles and the labium. Their form varies widely between different species.

Mechano-receptor   an organ sensitive to mechanical excitations such as vibrations, air and water currents and touch. In arthropods they often take the form of modified hairs which project from the body surface. They are particularly abundant on the antennae and legs.

Medusa   the aquatic, free-living, plate or bell-shaped sexually reproducing generation of a coelenterate. Jellyfish are large medusae.

Melanism   darkening due to accumulation of the pigment melanin. Industrial melanism is well-known. In the moth *Biston betularia* (no. 177) a dark form has replaced a light one on grime-covered lichenless trees in industrial England, possibly because the light ones are more conspicuous to predators.

Metamorphosis   transformation; the process of changing from the larva to the adult. Metamorphosis occurs in many different types of invertebrate as well as in some vertebrates, e.g. amphibians. In insects metamorphosis occurs in the Pterygota (see p. 178 for details) and is most complete in the Endopterygota where the larvae bear little resemblance to the adult. A special resting stage, the pupa, occurs in these insects between the last larval instar and the adult in which the larval organs and tissues are completely broken down and remodelled into those of the adult.

Migration   directed movement of individuals or of populations from one locality to another. Migrations are often seasonal occurrences.

Micro-climate   the climate within a restricted area, e.g. under a stone, on the undersurface of a leaf or in the top centimetre of the soil.

Micro-organisms   organisms visible only through the microscope, e.g. bacteria, protozoa etc.

Micropyle   minute hole in the shell of an insect egg through which sperm can penetrate.

Mimicry   a special form of protection in which one species comes to closely resemble another, often poisonous, distasteful or aggressive, species. Hover-flies (no. 46) may mimic bumble-bees (no. 40) and the wasp-beetle (no. 37) may mimic the common wasp (no. 38).

Mine   gallery or cavity within a living plant. Mines occur in leaves, stems, petioles etc. and are formed by the larvae of various insects including beetles (principally flea-beetles and weevils), sawflies, Lepidoptera and Diptera. Larvae within mines are well provided with food and live in conditions of favourable humidity and temperature.

Molluscicide   pest control chemical for killing molluscs.

Monophagy   a monophagous animal feeds on only one type of food; e.g. a single plant or animal species c.f. Polyphagy.

Monoculture   the cultivation of a single plant species in a district, as in a crop. Animals which feed on this particular species may multiply greatly and become pests.

Morphology   the study of the visible form of an organism.

Moulting   see Ecdysis.

Mutation   the sudden appearance of an alteration in an organism's hereditary make-up which is not due to re-combination of its genes. Mutations occur naturally but their incidence can be increased by exposure to certain chemicals and to radio-activity.

Mycelium   a mass of fungal strands.

Myrmecophile   an animal which lives in ant's nests. Many beetles, some parasitic wasps and even the larvae of certain butterflies (no. 24t) show this habit.

Nauplius larva   a type of aquatic larva with three pairs of appendages common amongst the Crustacea.

Nectar   a sugary, often scented liquid produced mainly by flowers. Many insects feed on nectar and bees convert it into honey. While collecting nectar insects carry pollen from one flower to another.

Nekton   general term for free-swimming animals which have sufficient powers of mobility as to be able to migrate freely and independently of water-currents cf. Plankton.

Nematocyst   the effective part of a nematoblast or stinging cell of a coelenterate. The nematocyst contains a long thread-like structure which at rest lies coiled within a capsule. On being stimulated mechanically or chemically the thread turns inside out and is shot out rapidly. It contains poison and can penetrate the skin of any animal coming into contact with it causing paralysis or death. The stings of jelly-fish are caused by nematocysts (see no. 384).

Nematicide   a poisonous chemical used for controlling nematodes.

Nematology   the study of nematodes.

Nomenclature   the naming of plants and animals.

Nymph   term often used for the larva of an exopterygote insect. Also the immature 8-legged stages of mites and ticks.

Obtect pupa a pupa (q.v.) in which the wings, legs etc. are firmly fixed to the body throughout their length so that they cannot be moved (e.g. no. 6, left).

Ocelli the simple eyes found on the top of the head of insects. The larvae of Endopterygota lack compound eyes but have simple eye-like structures in their place. The two types are distinguished as dorsal ocelli and lateral ocelli respectively.

Ommatidium the single element of a compound eye appearing on the surface as a facet.

Ovary the site of egg production in the female animal.

Oviparous descriptive of a species which lays eggs.

Ovipositor the egg-laying and egg-placing equipment of a female insect. Sometimes this appears as a series of elongated processes at the tip of the abdomen as in some hymenopterous parasites (e.g. no. 131), sometimes as a horny tip to the abdomen as in crane-flies (no. 134) and sometimes as a tube which is normally kept telescoped within itself inside the abdomen (as in beetles). The stings of bees and wasps are modified ovipositors which no longer have an egg-laying or placing function.

Ovoviviparous descriptive of an insect whose eggs hatch almost immediately after they have been laid.

Paedogenesis reproduction by an immature animal, e.g. a larva or a pupa. It is often, but not always associated with parthenogenesis (q.v.).

Palaeontology study of fossil animals and plants.

Palps sensory limb-like lateral processes on the maxillae and labium of insects and on some of the mouth-parts of other mandibulate arthropods. The ventral appendages of the prostomium (q.v.) of certain polychaete worms are also called palps.

Parasite an organism which lives in or on another organism (host) and from which it obtains its nourishment without destroying it or, in many instances, without causing it any severe harm. Ectoparasites live on the surface of their hosts, endoparasites within them. Warm-blooded animals are hosts both for ectoparasites (e.g. fleas and lice) and for endoparasites (e.g. some mites, warble-fly larvae, tape worms, and nematodes). Insects and other invertebrates also have ecto- and endoparasites (note the small red mites on the *Gerris lacustris* in no. 332) and they may act as vectors of parasites of warm-blooded species. (See also Parasitoid.)

Parasitoid a special form of lethal parasite found in insects and other arthropods. In a typical example eggs might be laid by the female of the parasitoid species into another insect. The parasitoid larva or larvae develop within their host and ultimately destroy it. Only the immature stages have the parasitoid habit, the adults are freeliving and usually feed elsewhere. Parasitoids themselves are affected by parasitoids, these are called secondary or hyperparasites. Parasitoids are most common in Hymenoptera and in the Tachinidae amongst the Diptera.

Paratype see Type.

Parthenogenesis virgin birth, a type of reproduction in which viable offspring are produced from unfertilised eggs. This is often combined with viviparity and alternation of generations (q.q.v.).

Pectinate with teeth like a comb (often applied to antennae).

Petiole the narrow waist of bees, wasps etc.

Pheromone a chemical substance produced by an insect which acts as a signal to other individuals of the same species. 'Sex pheromones' are attractive to the opposite sex and serve as allurements for mating (e.g. in Lepidoptera), 'trail pheromones' mark the tracks of ants to food or back to their nests and 'alarm pheromones' communicate a sense of danger within ant colonies. In the honey-bee the queen produces a pheromone which inhibits the building of queen-cells in the hive and suppresses the development of ovaries in the workers. During the last few years pheromones have been used for pest control. The sex pheromone of the female gypsy moth (larva no. 162) can be synthesised and used to attract the males which are then destroyed.

Photo-receptor an eye or other light-sensitive organ.

Phylogeny the evolution history of a species or group of animals or plants.

Physiology study of the life-processes of living organisms.

Phytophagous descriptive of a plant eating species.

Pigment a colour, e.g. melanin which forms in living cells.

Plankton the general term for small aquatic organisms which swim only feebly, if at all, and are thus carried helplessly by water currents c.f. Nekton.

Pleural relating to the sides of the body.

Pollen particles produced by the male reproductive organs of higher plants which contain the male germs. Pollen forms an important food of many animals, e.g. syrphid flies (nos. 46 − 50), pollen beetles (no. 58) and various bees.

Pollen basket   certain Hymenoptera, e.g. honeybees and bumblebees have an arrangement of hairs on the outside of their hind tibiae which serves to store collected pollen. This is the pollen basket.

Pollination   the transference of pollen from the stamen to the pistil so that the male and female germs can unite. Insects play a very important role in this and many flowers are specially adapted to attract them.

Polyembryony   a type of asexual (sexless) reproduction in which a single egg divides to give rise to many individuals — sometimes thousands. The phenomenon frequently arises amongst the hymenopterous parasitoids (see Parasitoid).

Polymorphism   the appearance of several different forms within an animal species. Sometimes these forms differ genetically, sometimes they do not and their appearance is controlled by nutritive and other external factors. Often the forms are adapted for certain functions (functional polymorphism) as in termites where queens, males, soldiers and workers occur and have separate duties.

Polyp   the sessile generation of a coelenterate. Typically this reproduces asexually by budding to produce the sexual form, a medusa (q.v.).

Population   a group of individuals of a species or sub-species within a geographically or ecologically defined region.

Posterior   the hindmost, e.g. posterior wings (c.f. Anterior).

Predator   an animal which catches, kills and wholly or partially eats other animals.

Primitive   not specialised, an animal or plant which is at an early stage of evolutionary development.

Proboscis   the projecting tubular mouthparts of certain insects e.g. Lepidoptera.

Pronotum   the dorsal part of the first thoracic segment in insects. It is particularly well-developed in Orthoptera, Dictyoptera, Coleoptera and certain Hemiptera.

Prolegs   the unjointed abdominal limbs of caterpillars.

Prostomium   the pre-segmental part of an annelid worm.

Protective colouration   camouflage, mimicry of colours etc. (q.q.v.).

Protective resemblance   mimicry and resemblance to environment in form and structure as well as in colour, e.g. larva in no. 180 resembles a twig, bug in no. 203 resembles a thorn. The phenomenon is common in insects, spiders etc.

Prothorax   the anterior thoracic segment of an insect.

Pterostigma   a coloured spot near the tip of the wing of a dragonfly (no. 300) or at the anterior margin of the forewing of many Hymenoptera.

Pterygota   the winged, or secondarily wingless insects (see p. 178).

Pupa   the instar which follows the larval stages of an endopterygote insect. Within the pupa the tissues of the larva are broken down and remodelled into those of the adult. A pupa may be enclosed within a silken cocoon or it may be supported by a silken girdle. The pupae of cyclorrhaphan flies are enclosed within a puparium — the hardened last larval skin. See also Exarate pupa and Obtect pupa.

Pupiparous   descriptive of an insect which gives birth to full-grown larvae ready to pupate. Some parasitic Diptera show this phenomenon.

Queen   the reproductive female of a social insect, e.g. ant, bee, wasp or termite.

Radula   the file-like tongue found in the mouths of gasteropod, scaphopod and cephalopod molluscs.

Race   see sub-species.

Recent   relating to the present geological epoch.

Receptor   organ which receives a stimulus see Chemo-, Mechano- and Photo-receptor.

Reflex   uncontrolled stereotyped reaction to a stimulus, e.g. blinking when something crosses the eye.

Regeneration   the reformation of tissues, organs and other lost body parts. Many simple animals such as flat-worms, starfishes and sea-squirts show considerable powers of regeneration and so do some amphibians and reptiles. In insects and other arthropods regeneration occurs only in connection with moulting (see Autotomy).

Relict   surviving remnant, within an area, of a group which was earlier more widely distributed.

Resistance   opposition; lack of susceptibility. Many insect populations show resistance to certain insecticides because after repeated exposure to them the more susceptible individuals have been eliminated so that now the less susceptible ones make up the whole population.

Respiration   breathing.

Raptorial   a predator is raptorial if it is adapted for seizing and holding its prey. The 'mask' of a dragon-fly larva is a raptorial structure.

Rostrum   the projecting tubular mouthparts

of a hemipteran or the snout of a weevil.

Scape the first segment of the antenna of an insect. In many species the scape is elongated and the antenna has an 'elbowed' appearance as in the pine weevil (no. 226).

Scarabaeiform a curved, soft-bodied, long legged larva as found in the family of beetles, Scarabaeidae (no. 272).

Sclerite a hardened plate forming part of the exoskeleton of an arthropod (see Cuticle).

Scutellum a usually triangular region between the fore-wing bases of some insects. Well seen in beetles and Heteroptera.

Secretion a chemical substance or mixture of substances produced and emitted by a gland. E.g. Saliva is a secretion and so are pheromones (q.v.) emitted by insects.

Segment bodies of arthropods, annelids and vertebrates are fundamentally built up of consecutive, basically similar regions, the segments. In primitive forms these appear as separate body rings (e.g. worm no. 284) but in more advanced forms the ringed appearance is lost and the segments differ greatly. The term is also used for the consecutive parts of arthropod appendages, e.g. segments of the antennae, legs etc.

Selection choice. Natural selection is one of the main mechanisms of evolutionary change. All organisms over-produce offspring and there is considerable variation between individuals. Individuals less well adapted for life tend to be weeded out selectively before they can breed so that the better adapted ones become the parents of the next generation. Hence the population gradually evolves towards improved adaptation.

Serrate toothed like a saw, e.g. serrate antennae (no. 111).

Sessile a sessile animal is essentially non-mobile, e.g. sea anemone or sea-mat.

Seta a bristle of an insect or mite.

Silk a proteinaceous substance produced as threads from the silk glands of many types of insect and arachnid. Silk is often used to make a cocoon (as in the silk-worm moth), it may be used to spin leaves together (certain caterpillars) and spiders make their webs from it.

Silk-glands these produce silk and are found on various parts of the body of arthropods. Caterpillars have them on the labium and spiders have them on the opisthosoma where they open on spinnerets (p. 177).

Siphon a tubular extension of the mantle-cavity in many bivalve, cephalopod and prosobranch molluscs which serves to circulate water over the gills. The elongated caudal breathing spines of aquatic bugs such as waterscorpions (no. 337) are also termed siphons.

Social insects insects such as ants, bees, wasps and termites which live together in organised communities. The duties of the whole community are shared between individuals. e.g. in the honey-bee the queen lays eggs, the drones fertilise the queens and the workers build the combs, feed the young and collect food for the community. Different ages of workers have different duties, e.g. the youngest clean the hive, the eldest forage for food.

Solitary insects insects which do not organise into communities. The term is usually reserved for solitary bees and wasps to distinguish them from the social ones.

Sound-production animals may produce sound as a by-product of their activity e.g. the buzz of flies is produced by the vibration of the wings. Many insects however have special sound-producing organs and these occur particularly in the males where they serve to attract the females. In bush-crickets and crickets sounds are made by rubbing the bases of the fore-wings together, grasshoppers rub their spiny back-legs against the fore-wings or body-wall and cicadas have drum-like organs on the abdomen with which they make a noise which has been likened to the whistle of a railway engine. Many beetles produce sound and here its function may be defensive or it may assist in keeping individuals of a population together. Sound produced by rubbing two parts of the body together is termed stridulation.

Species the basic unit in the classification of animals and plants (see p. 156). A species includes all individuals of similar structure which are capable of reproducing to produce viable offspring. As a rule different species do not cross and in instances where crossing does occur, the offspring, termed hybrids, are usually sterile or of greatly reduced fertility. When populations of a species become isolated from each other they tend to evolve in slightly different directions so that sub-species arise. Breeding between sub-species can occur and if their populations meet there tends to be a gradual merging of one form into the other across the territory they occupy. Systematists, often working with dead animals only, find it difficult to decide which forms are sub-species and which are true species.

Species-specific describes an event or occurrence relating to one species only. e.g. some parasitoids attack only one species of host.

Sperm the motile cells containing the male germ.

Spermatophore in certain animals the sperm is transferred to the female encapsulated in a receptacle of dried secretion termed a spermatophore. Spermatophores occur in various crustaceans, insects, arachnids and molluscs.

Spinnerets papillae on the hind-body of spiders upon which the openings of the silk-glands occur (see p. 177 and Silk-glands).

Spiracle breathing-hole of tracheate arthropod. In insects and some other forms the spiracles are segmentally arranged and appear as small holes along the sides.

Sterile infertile, incapable of reproducing.

Sternite the ventral plate of an arthropod body segment.

Stigma another term for spiracle especially with reference to mites. Also the pigment spot near the photo-receptor of a flagellate.

Stimulus something which irritates or excites a sensory receptor of an animal and causes a response.

Sting the modified ovipositor of Hymenoptera of the sub-order Apocrita. Stings are associated with poison glands and are used by parasitoids to paralyse or kill their prey and by ants, bees and wasps, where their egg-laying function has been lost, mainly as weapons of offence and defence.

Stink-glands glands which produce foul substances probably serving as means of defence. They are found in Hemiptera such as pentatomids, in millipedes and in harvestmen.

Stomach poison a biocide (q.v.) which affects the animal only after it has been taken into the alimentary canal.

Stridulation many insects, e.g. grasshoppers, can produce sound by rubbing two hard surfaces together. This is stridulation (see Sound-production).

Stylets needle like appendages of arthropods. Many insects and mites have mouthparts adapted for piercing and sucking tissues. The needle-like parts of these are termed stylets. Needle like ovipositor components in Hymenoptera are also called stylets.

Sub-imago the fully winged but pre-adult penultimate instar of a mayfly. This stage only lasts for a few minutes in some species but persists for as long as two days in others. This is the 'dun' of the angler.

Sub-species see Species.

Symbiosis an association of two different species which benefits both. e.g. ants may tend and protect aphids and place them on suitable stems or roots, in return the aphids produce honeydew which is imbibed by the ants. Another example is a sea-anemone carried on the shell inhabited by a hermit-crab. The anemone protects the crab, the crab helps the anemone get food. Symbionts are usually dependent on each other. (c.f. Commensalism, Parasite).

Synonym animals and plants are often named inadvertently more than once. The various names for one particular species are synonyms and the earliest name is taken to be the valid one (c.f. Homonym).

Systematics the branch of biological study in which the worker arranges animals and plants into various groups preferably according to their evolutionary relationships (see pp. 156 – 176).

Tactile relating to the sense of touch. (See Mechano-receptor.)

Tarsus the foot of an insect leg.

Taxon any systematic unit e.g. species, genus, family, phylum.

Taxonomy scientific study of classification of animals and plants.

Temperature-sense not much is known of the temperature-senses of primitive animals but insects have well developed means of appreciating temperature differences. The receptive organs are often placed on the antennae (many blood-sucking forms) or on the feet (cockroaches) but may be distributed over the entire body (grasshoppers). Some insects can orientate themselves to pick up radiant heat e.g. some pentatomids at low temperatures direct their backs towards a source of light, presumably to acquire warmth.

Tentacle tactile feeler, arm or other elongate body appendage in animals such as polychaetes, gastropods, cephalopods, coelenterates etc. Tentacles may carry sensory receptors or they may be used for capturing and holding prey.

Thermophile a lover of warmth.

Terrestrial relating to or living on land.

Thorax the middle region of an insectan or crustacean body bearing the jointed walking (or swimming) legs. In insects the region is well defined, and consists of three segments, in the crustaceans it is variable in extent.

Tibia the part of an insect leg between the femur and the tarsus.

Trachea the breathing tubes of insects, chilopods, diplopods, many arachnids and other tracheate arthropods. The tracheae are branching structures lined with cuticle which is cast off at every moult. They communicate to the outside through spiracles and at their inner extremities divide into fine tubules which penetrate between and into cells so that oxygen can be brought directly to the tissues. In these animals the blood plays little part in the transportation of oxygen. Air-sacs (q.v.) are often associated with the tracheae. (See Tracheal gills).

Tracheal gills thin-walled out-growths of the body wall richly supplied with fine branches of tracheae. These occur in the larvae of various aquatic insects e.g. dragonflies, mayflies, stoneflies and caddisflies. Oxygen from the water passes through the walls of these gills and enters the tracheae which usually form a closed system.

Trochanter the small joint between the coxa and femur of an insect leg.

Tympanal organs the well-developed hearing organs of Orthoptera, cicadas, Hemiptera such as *Sigara* and Lepidoptera. These organs usually consist of a thin membrane of cuticle, the 'tympanum' which is associated with tracheal air-sacs and sensory receptors. They may be sensitive to a wide range of sound frequencies, even to ultra-sonics.

Type the description of a species is usually based on a series of specimens which constitute the type series. These are kept for reference and one specimen, the holotype, is labelled as such and regarded as the ultimate resort for settling any future problems of identification. (Type-series have on many occasions been proved later to be mixtures of species). A single individual of the opposite sex to the holotype is the allotype. All specimens except the holotype within the type series are paratypes.

Ultra-sonics sound waves with frequencies too great to be heard by man. Young people can hear sounds with frequencies of up to about 20,000 Hz. but for older people the limit is appreciably lower. Some Orthoptera, however, can produce and respond to sound-frequencies of around 90,000 Hz. A considerable part of their song is inaudible to us and old people may hear nothing at all. Bats emit ultra-sonics with frequencies of up to 100,000 Hz. and use them as a kind of radar. Certain moths can detect these sounds at a distance of 100 feet or more.

Ultra-violet light radiation of shorter wavelength than violet light and invisible to man. Many insects such as bees and blowflies see ultra-violet as colours and, as can

be shown by ultra-violet photography, they see patterns in flowers that are quite invisible to us.

Vector a bearer. Many insects are vectors of disease organisms. (See Disease vectors.)

Ventral relating to the underside (c.f. Dorsal).

Vertebrate an animal such as a fish, amphibian, reptile, bird or mammal which possesses a backbone.

Viviparity production of living young rather than of eggs.

Warning colouration insects with effective methods for defence are often conspicuously and characteristically patterned with bright colours such as red, orange and yellow. This warns would-be predators of their dangerous or unpleasant nature. Examples are wasps which can sting, burnet moths which are poisonous and ladybirds which are foul-smelling.

Wax secretion occurs in some aphids, coccids and sawfly larvae and serves to protect their bodies. Bees secrete wax and this is used to build up the cells of the honeycomb.

Wings among invertebrates only insects have wings. Not all insects are winged. The Apterygota have none and their wingless condition is primitive i.e. their ancestors were wingless also. Certain Pterygota, e.g. fleas and lice, are wingless but this condition is secondary i.e. they are descendants of winged insects but the wings have become lost during their evolutionary development. The wings only appear fully formed in the adult insect. In the Exopterygota they can be seen as external wing pads which increase in size from instar to instar. In the Endopterygota they develop internally and appear externally in the pupa. The wings are found on the second and third thoracic segments only; in certain insects they become modified e.g. the fore-wings to shell-like elytra in the beetles.

Wing-dimorphism this occurs when certain individuals of a species have fully developed wings whereas others have abbreviated wings or none at all. The phenomenon occurs in some grasshoppers, moths, beetles, bugs, stoneflies, ants, and aphids and is often combined with sexual dimorphism. Nutritional factors may also be involved – some parasitoids reared on meagre food supplies fail to develop their wings properly. Wing-polymorphism also occurs.

Xerophile a species which likes dry conditions and avoids damp ones.

Index

This index refers to the colour plates, their titles and their explanatory texts only.

A figure by itself indicates that the entry listed is the main subject of the plate with that number.

A figure followed by t indicates that the entry receives mention in the relevant text only.

A figure in brackets following the name of an animal group indicates that a representative of this group is mentioned in title or text but the group name itself is not.